FRCR Part 1:
Cases for the Anatomy
Viewing Paper

FRCR Part 1:
Cases for the Anatomy
Viewing Paper

Dr James D. Thomas

Editorial advisors
Dr Christopher Fang
Dr Udara Kularatne

OXFORD
UNIVERSITY PRESS

OXFORD
UNIVERSITY PRESS

Great Clarendon Street, Oxford ox2 6DP

Oxford University Press is a department of the University of Oxford.
It furthers the University's objective of excellence in research, scholarship,
and education by publishing worldwide in

Oxford New York

Auckland Cape Town Dar es Salaam Hong Kong Karachi
Kuala Lumpur Madrid Melbourne Mexico City Nairobi
New Delhi Shanghai Taipei Toronto

With offices in

Argentina Austria Brazil Chile Czech Republic France Greece
Guatemala Hungary Italy Japan Poland Portugal Singapore
South Korea Switzerland Thailand Turkey Ukraine Vietnam

Oxford is a registered trade mark of Oxford University Press
in the UK and in certain other countries

Published in the United States
by Oxford University Press Inc., New York

British Library Cataloguing in Publication Data
Data available

Library of Congress Cataloging in Publication Data
Data available

Typeset by Cenveo, Bangalore, India
Printed in Great Britain
on acid-free paper by
CPI Group (UK) Ltd, Croydon, CR0 4YY

ISBN 978–0–19–960453–1

10 9 8 7 6 5 4 3 2 1

FOREWORD

Anatomy is a core fundamental to the practice of radiology. A thorough knowledge of anatomy is the key to describing the site of disease or abnormality, and often in understanding a disease process and its complications. When I commenced my radiology training in 2002, anatomy had been removed from the FRCR Part 1 syllabus. Although its exclusion from the syllabus seemed slightly odd at the time, it did mean less revision and one less thing to fail. In the end, I never did avoid that exam in anatomy. Perhaps appealing to their sadistic instincts, my trainers still subjected us to an 'old style' Part 1 examination at the end of our first year. Looking back, this was certainly of no detriment to my training or subsequent working career.

Anatomy now makes a return to the Part 1 examination. I have been fortunate to have been involved with this book. The author has made great efforts to provide you with variation between the papers, with both straightforward and more challenging questions. To ensure clarity and accuracy, the papers have been prepared meticulously with input from those who have sat the new examination and tested out on other trainees.

In helping with the preparation of this book, I have found the questions outside my subspecialty interest both challenging and a source of revision. I am therefore confident that, for the intended audience, it will prove to be a useful resource to tighten up areas of weakness and encourage further reading around the answer.

My training and career have taught me that a radiologist can never know too much anatomy.

Good luck.

Dr Christopher Fang
Consultant Radiologist
Royal Derby Hospital

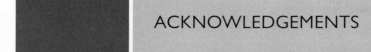

ACKNOWLEDGEMENTS

I would like to extend my gratitude to Dr Chris Fang who tirelessly read all the exams and answers, pointing out inaccuracies and making suggestions for improvement. This book has benefited greatly from his involvement.

Thanks also goes to Dr Udara Kularatne who contributed the tricky examinations 7 and 8. He also read through the other examinations and provided a constant stream of feedback and ideas. His enthusiasm is infectious.

I would also like to thank the registrars who road-tested the examinations: Dr Abdullah Saeed, Dr Waleed Al Obaydi, Dr Sachin Srivastava, Dr Arieff Abu Hassan, Dr Amit Bharath, and Dr Walter Sprenger de Rover.

My thanks also goes to Chris Reid and others at Oxford University Press for bringing this from idea to reality.

Finally, extra-special thanks to my other half, Tanya Monaghan, for putting up with me, this project, and the mountains of paper that now fill half the house.

Dr James D. Thomas
Specialist Registrar in Radiology
Nottingham University Hospitals NHS Trust

The author and the publisher would like to thank Dr Antoine Rosset for permission to use screenshots of the OsiriX DICOM viewer.

CONTENTS

INTRODUCTION

The examination structure

The anatomy paper consists of 20 images (called 'cases'), with 5 questions each. The images contain arrows with alphabetical labels, usually from A to E, although some may contain fewer depending on the question structure.

You will have 75 minutes to answer the questions.

At the time of writing, the images/cases are shown digitally. Each candidate has their own computer workstation and these are arranged in rows with plenty of space between candidates. Each workstation consists of a 19" monitor with an Apple Mac mini® attached. There is a mouse, but no keyboard. The mouse will either be a standard Apple mouse, or an Apple 'magic mouse'®. The right click function on the mouse is disabled.

The Apple 'magic mouse'® has no visible buttons or scroll wheel. The surface is touch-sensitive; left and right click are performed in the usual way on the left and right sides of the mouse and scrolling is performed by wiping the surface of the mouse with a finger as if using a touchpad. Candidates for the first part of the FRCR will not need to use either of these functions so should not be put off if unfamiliar with this.

The questions are presented in a booklet, in which you also write your answers. The choice of pen or pencil is not specified and you will need to bring your own. Using a pencil, however, would allow corrections to be made easily. If using a pencil, remember to bring an eraser and a sharpener.

OsiriX image-viewing software

The images/cases are shown using 'OsiriX', a free Apple-based DICOM- and image-viewing program developed and maintained by Pixmeo, a Swiss company based in Geneva. Before the examination begins, the invigilators will give a short tutorial on using the software and you will have some example images to practice with. Apple users may wish to download this from the OsiriX website (www.osirix-viewer.com) to familiarize themselves with the tools. The software is easy to use and non-Mac users will not be disadvantaged.

Apple windows

Candidates unfamiliar with Apple software will notice the brightly coloured circles at the top left of each window (Figure 1). These are red, amber, and green. These act to close, minimize, and optimize the window size, respectively. DO NOT press any of these during the examination.

The main image-viewing window

The images occupy the majority of the screen and the central window. To the left, there is a list of the images in the series. See Figure 1 for an example. The list on the left can be used to navigate through the images if necessary.

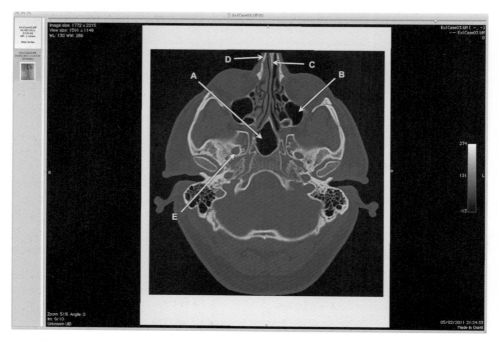

Figure 1

The top of the window shows the file name of the image being displayed. You should double-check this when entering your answers to make sure that you are filling in the correct question numbers.

The top toolbar

The OsiriX software has several image manipulation controls in the top toolbar as shown in Figure 2. The majority of these are either removed or disabled for the purposes of this examination. In particular, you should ignore anything related to window presets, 2D/3D, orientation, slab thickness, and the 'title'. Some of these are used in the Part 2B examination but none are needed for the Part 1.

Figure 2

The controls that may be of use to you are:

- The 'Database' button: used to return to the list of images/ cases to navigate through the examination set.
- The left and right arrows: used to advance to the next case or return to the previous case (the author recommends using these to proceed through the examination and using the 'Database' button to return to specific images at the end if necessary).

- The 'mouse button function' buttons: used to manipulate the image. A close-up is shown in Figure 3.

The mouse button functions are the only tools that you might need during this examination and are shown in Figure 3. From left to right they are:

- Window level: click this to activate the window levelling controls. Clicking and dragging left/ right on the image adjusts the window width; up/ down adjusts the window level. This should be familiar to all PACS users.
- Move: use this tool to drag the image around the screen (used if you have zoomed in).
- Zoom: once this is active, click and drag upwards to zoom in, click and drag down the screen to zoom out. You can then look around the zoomed-in image using the 'move' tool, which you must select from the toolbar (see above).
- Rotate: this allows you to swivel the image around a central point. You should not need to use this.
- Animate: this is used to scroll through stacked images (such as a CT examination). You will not need this for the first part of the FRCR examination.
- ROI tool: this symbol looks like a green line. Activating this and clicking on the image will draw green lines with length measurements. These are impossible to delete without a keyboard. DO NOT press this.

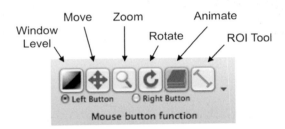

Figure 3

General advice for the examination

Read the question!

Although most of the questions will be of the nature 'Name the structure labelled ...', some may not be. Be sure to read the question carefully each time to avoid being caught out and needlessly losing a mark.

Left or right?

Candidates should write the side of the structure if possible. For example, if a radiograph of a thigh is clearly marked with an 'R', the bone is the 'right femur'. Axial CT (and MRI) images will be presented in the conventional way with the patient's right on your left as you view the screen and vice versa. Again, the side of the structure marked should be written in the answer.

In some cases, the side will not be obvious. There may be a radiograph or cross-sectional image of a limb with no side marker, for example. The candidate would be wise to write 'no side marker' to indicate that they have tried and failed to determine the side.

Is that the 'femur' or the 'greater trochanter'?

Confusion may arise if an arrow appears to point to part of a structure. If several parts of the same structure are labelled, then the arrows clearly indicate those individual parts.

In general, the candidate would be wise to name the structure in the most accurate way possible. An arrow to the greater trochanter, therefore, should be answered as 'greater trochanter of the femur'. If the examiners were only looking for 'femur', you're not going to lose any marks! Of course, there is a small risk that you name the wrong part and lose a mark that way.

Attempt every question

There is no negative marking so there is nothing to be gained from leaving an answer blank. Only you and the marking examiners will know if you gave a stupid wrong answer, so take the risk!

Write clearly

The examination is hand-marked by two examiners, so ensure you write clearly for the examiner to read your answer easily. If you have particularly illegible handwriting, consider writing in capital letters, as long as this does not slow you down.

Good luck!

You have 75 minutes to complete the examination

Difficulty rating: Ben Nevis

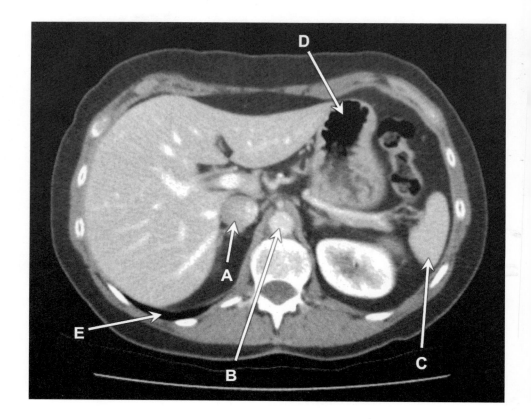

Case 1

1. Name the structure labelled A.
2. Name the structure labelled B.
3. Name the structure labelled C.
4. Name the structure labelled D.
5. Name the structure labelled E.

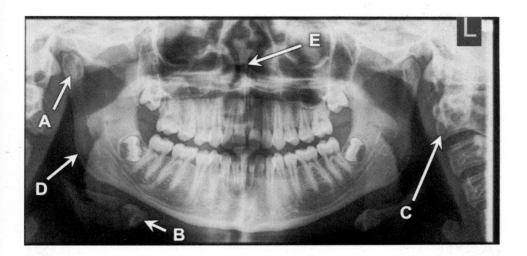

Case 2

1. Name the structure labelled A.
2. Name the structure labelled B.
3. Name the structure labelled C.
4. Name the structure labelled D.
5. Name the structure labelled E.

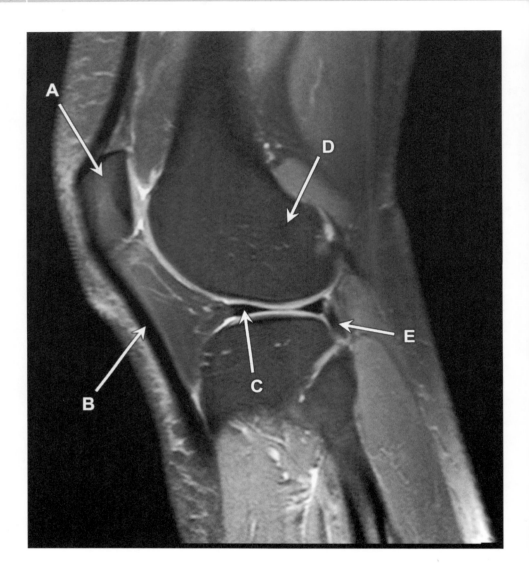

Case 3

1. Name the structure labelled A.
2. Name the structure labelled B.
3. Name the structure labelled C.
4. Name the structure labelled D.
5. Name the structure labelled E.

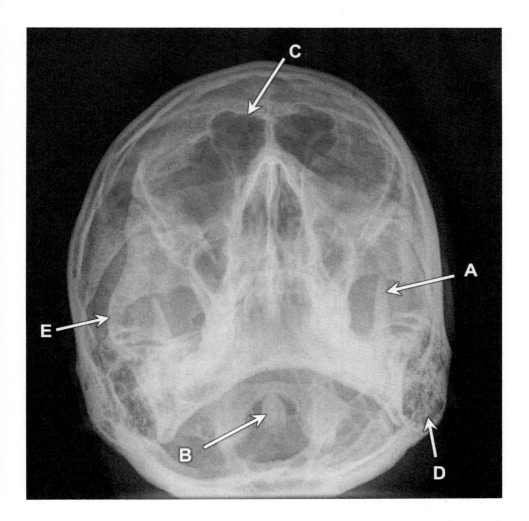

Case 4

1. Name the structure labelled A.
2. Name the structure labelled B.
3. Name the structure labelled C.
4. Name the structures labelled D.
5. Name the structure labelled E.

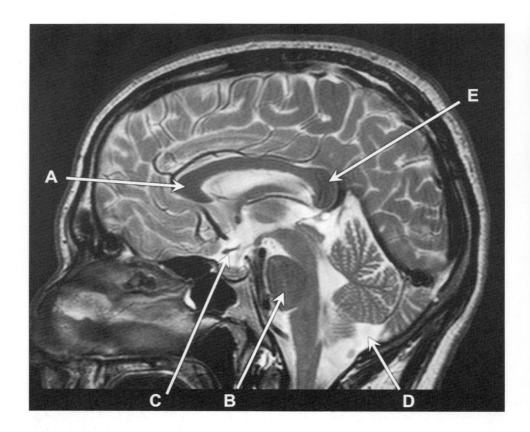

Case 5

1. Name the structure labelled A.
2. Name the structure labelled B.
3. Name the structure labelled C.
4. Name the structure labelled D.
5. Name the structure labelled E.

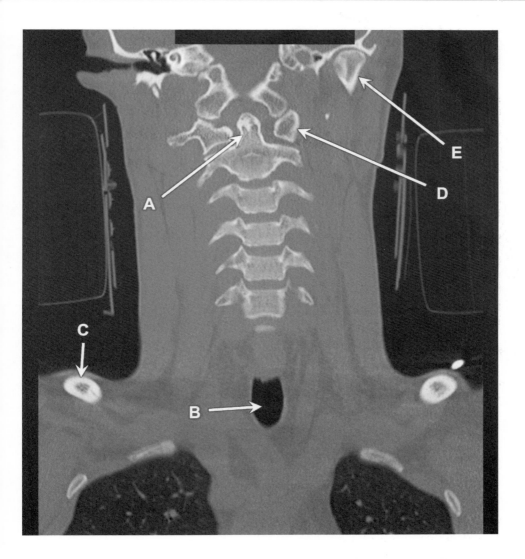

Case 6

1. Name the structure labelled A.

2. Name the structure labelled B.

3. Name the structure labelled C.

4. Name the structure labelled D.

5. Name the structure labelled E.

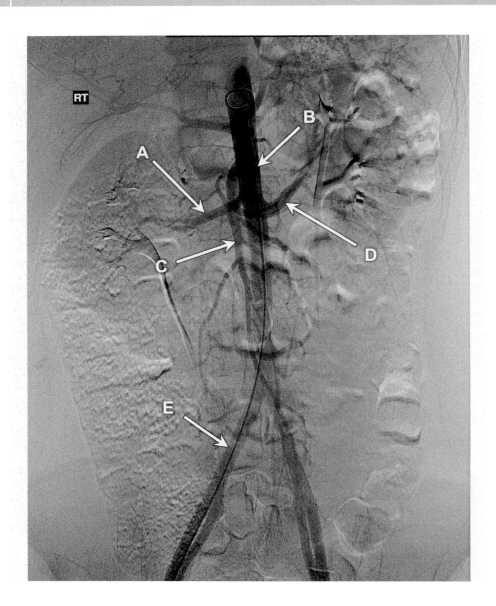

Case 7

1. Name the structure labelled A.
2. Name the structure labelled B.
3. Name the structure labelled C.
4. Name the structure labelled D.
5. Name the structure labelled E.

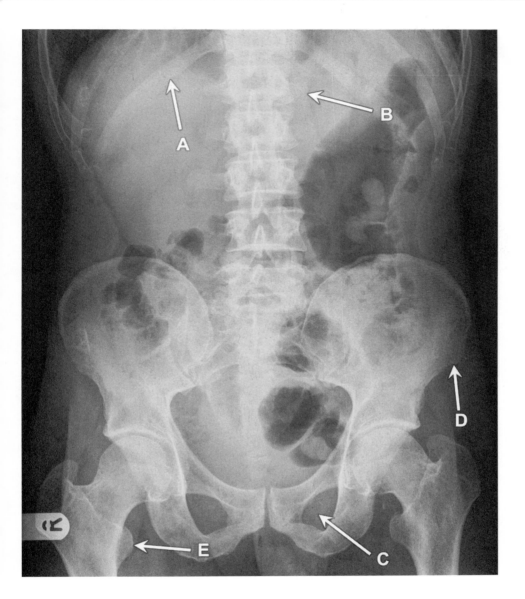

Case 8

1. Name the structure labelled A.

2. Name the structure labelled B.

3. Name the foramen labelled C.

4. Name the muscle that attaches to the structure labelled D.

5. Name the structure labelled E.

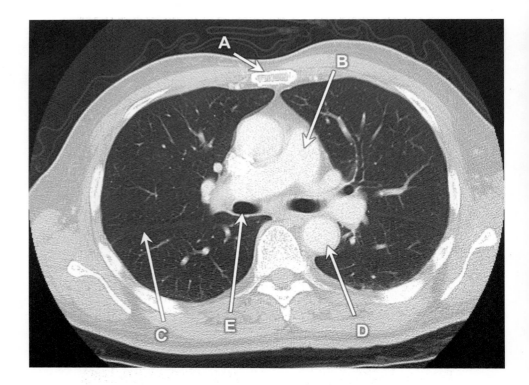

Case 9

1. Name the structure labelled A.
2. Name the structure labelled B.
3. Name the structure labelled C.
4. Name the structure labelled D.
5. Name the structure labelled E.

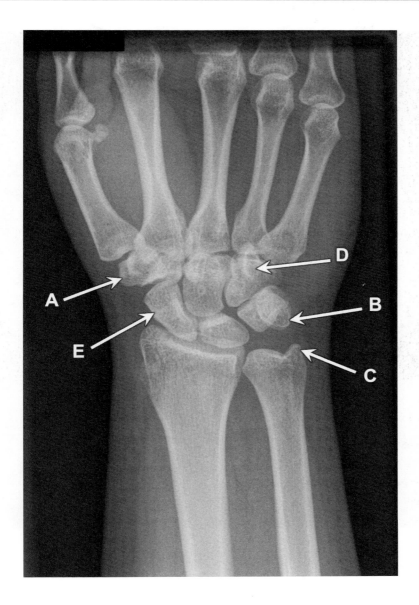

Case 10

1. Name the structure labelled A.
2. Name the structure labelled B.
3. Name the structure labelled C.
4. Name the structure labelled D.
5. Name the structure labelled E.

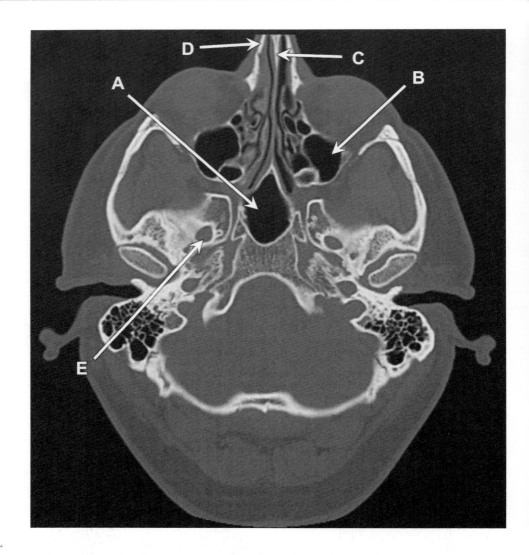

Case 11

1. Name the structure labelled A.
2. Name the structure labelled B.
3. Name the structure labelled C.
4. Name the bone labelled D.
5. Name the structure labelled E.

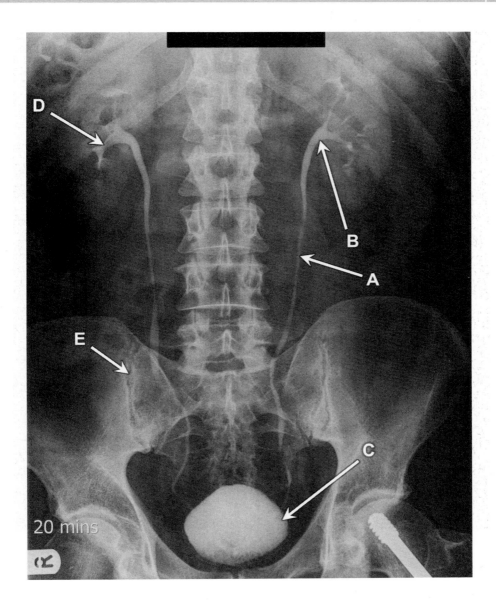

Case 12

1. Name the structure labelled A.
2. Name the structure labelled B.
3. Name the structure labelled C.
4. Name the structure labelled D.
5. Name the structure labelled E.

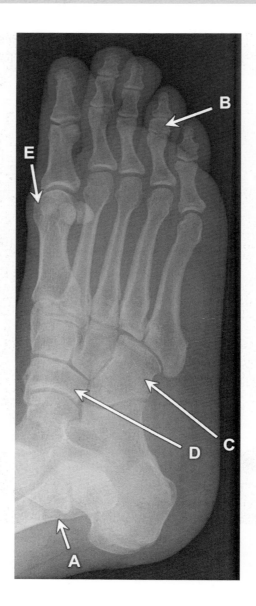

Case 13

1. Name the structure labelled A.

2. Name the structure labelled B.

3. Name the structure labelled C.

4. Name the structure labelled D.

5. Name the structure labelled E.

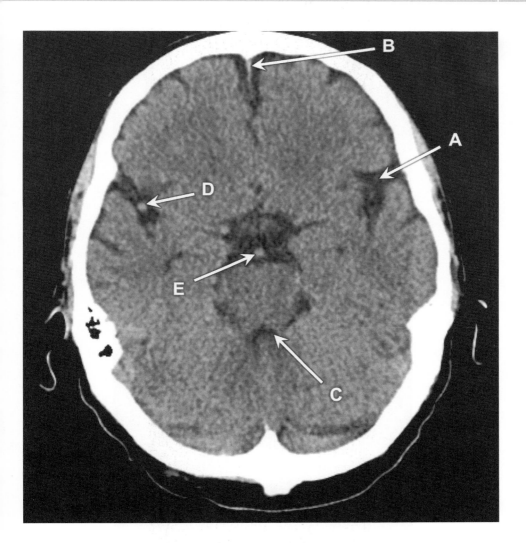

Case 14

1. Name the CSF space labelled A.
2. Name the structure labelled B.
3. Name the CSF space labelled C.
4. Name the structure labelled D.
5. Name the structure labelled E.

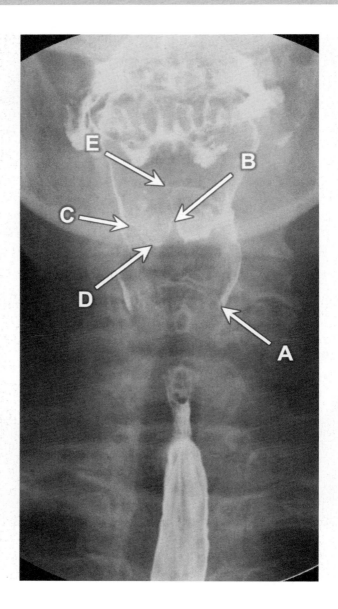

Case 15

1. Name the structure labelled A.

2. Name the structure labelled B.

3. Name the structure labelled C.

4. Name the structure labelled D.

5. Name the structure labelled E.

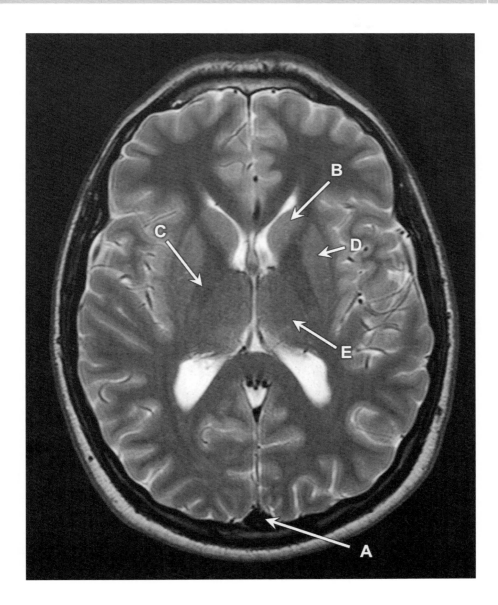

Case 16

1. Name the structure labelled A.
2. Name the structure labelled B.
3. Name the structure labelled C.
4. Name the structure labelled D.
5. Name the structure labelled E.

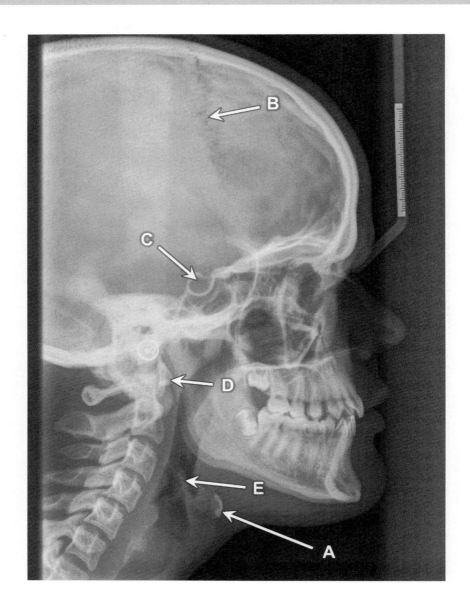

Case 17

1. Name the structure labelled A.
2. Name the structure labelled B.
3. Name the structure labelled C.
4. Name the structure labelled D.
5. Name the soft tissue opacity labelled E.

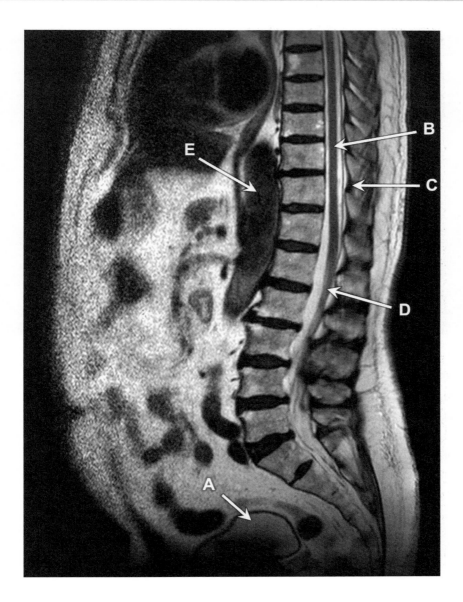

Case 18

1. Name the structure labelled A.

2. Name the structure labelled B.

3. Name the structure labelled C.

4. Name the structure labelled D.

5. Name the structure labelled E.

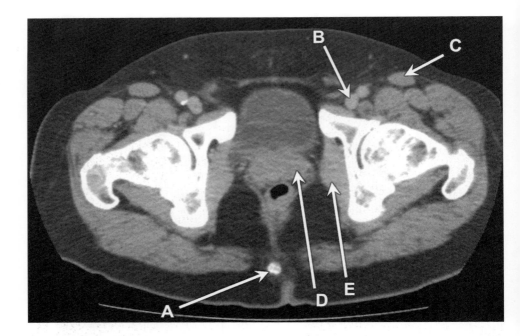

Case 19

1. Name the structure labelled A.

2. Name the structure labelled B.

3. Name the structure labelled C.

4. Name the structure labelled D.

5. Name the structure labelled E.

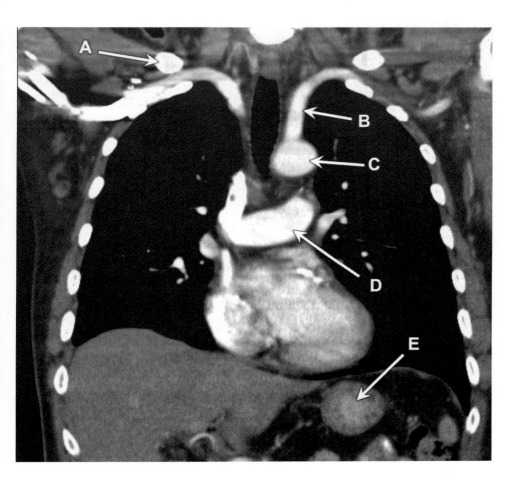

Case 20

1. Name the structure labelled A.

2. Name the structure labelled B.

3. Name the structure labelled C.

4. Name the structure labelled D.

5. Name the structure labelled E.

Case 1

CT abdomen. Axial section.

1. Inferior vena cava
2. Abdominal aorta
3. Spleen
4. Stomach
5. Right lung base (lower lobe)

Case 2

Orthopantomogram.

1. Right mandibular condyle
2. Hyoid bone
3. C2 vertebral body
4. Angle of the mandible (right)
5. Nasal septum

Case 3

MRI knee. T2W (with fat-suppression). Sagittal section.

1. Patella
2. Patellar tendon
3. Anterior horn of the lateral meniscus
4. Lateral femoral condyle
5. Popliteus tendon

As the fibula is included in the image, this must be the lateral part of the knee.

Case 4

Occipitomental radiograph of skull.

1. Coronoid process of the mandible
2. Dens/ odontoid peg
3. Right frontal sinus
4. Left mastoid air cells
5. Right zygomatic arch

Case 5

MRI brain. T2W sagittal section.

1. Genu of the corpus callosum
2. Pons
3. Optic chiasm
4. Cisterna magna
5. Splenium of the corpus callosum

Case 6

CT C-spine. Coronal section.

1. Dens/ odontoid peg
2. Trachea
3. Right clavicle
4. Left lateral mass of CI vertebra
5. Left mandibular condyle

Case 7

Abdominal aortogram (fluoroscopic).

1. Right renal artery
2. Abdominal aorta
3. Superior mesenteric artery
4. Left renal artery
5. Right common iliac artery

Case 8

Plain radiograph. AP abdomen.

1. Right twelfth rib
2. Left transverse process of LI vertebra
3. Left obturator foramen
4. Left sartorius
5. Right lesser trochanter

Case 9

CT chest. Axial section.

1. Sternum
2. Pulmonary trunk
3. Right oblique fissure
4. Descending thoracic aorta
5. Right main bronchus

Case 10

Plain radiograph. DP wrist.

1. Trapezium
2. Pisiform
3. Ulnar styloid process
4. Hook of the hamate
5. Scaphoid

Case 11

CT head. Axial section.

1. Sphenoid sinus
2. Left maxillary sinus
3. Nasal septum
4. Right nasal bone
5. Right foramen ovale

Case 12

Intravenous urogram, 20-minute radiograph.

1. Left ureter
2. Left renal pelvis
3. Urinary bladder
4. Right lower pole (major) calyx
5. Right sacroiliac joint

Case 13

Plain radiograph. Oblique foot.

1. Fibula
2. Middle phalanx of fourth toe
3. Cuboid
4. Navicular
5. First metatarsal

Case 14

CT head. Axial section.

1. Left Sylvian fissure
2. Falx cerebri (anterior part)
3. Fourth ventricle
4. Right middle cerebral artery
5. Basilar artery

Case 15

Barium swallow. AP pharyngeal view.

1. Left pyriform sinus
2. Median epiglottic fold
3. Right lateral epiglottic fold
4. Right vallecula
5. Epiglottis

Case 16

MRI brain. T2W axial section.

1. Superior sagittal sinus
2. Head of left caudate nucleus
3. Posterior limb of right internal capsule
4. Left lentiform nucleus
5. Left thalamus

Case 17

Encephalogram.

1. Hyoid bone
2. Coronal suture
3. Pituitary fossa
4. Anterior arch of C1 vertebra
5. Epiglottis

Case 18

MRI lumbar spine. T2W sagittal section.

1. Urinary bladder
2. Spinal cord
3. Ligamentum flavum
4. Cauda equina
5. Abdominal aorta

Case 19

CT pelvis. Axial section.

1. Coccyx
2. Left common femoral vein
3. Left sartorius muscle
4. Prostate
5. Left obturator internus muscle

Case 20

CT-pulmonary arteriogram (CTPA). Coronal section.

1. Right clavicle
2. Left subclavian artery
3. Aortic arch
4. Pulmonary trunk
5. Stomach

You have 75 minutes to complete the examination

Difficulty rating: Ben Nevis

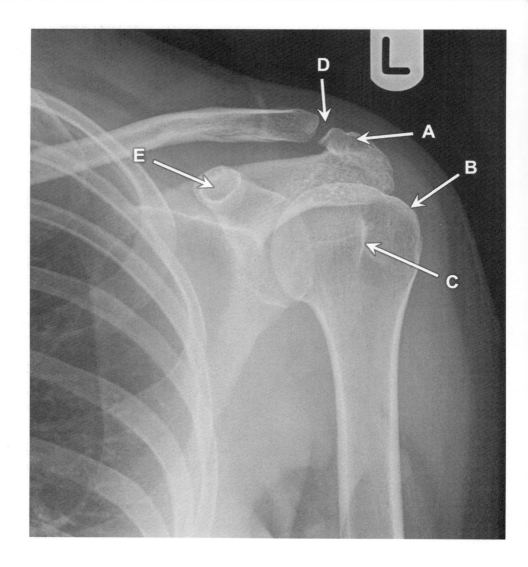

Case 1

1. Name the structure labelled A.
2. Name the structure labelled B.
3. Name the structure labelled C.
4. Name the structure labelled D.
5. Name the structure labelled E.

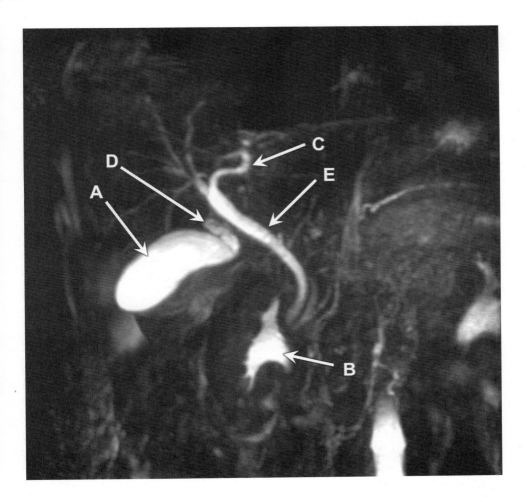

Case 2

1. Name the structure labelled A.
2. Name the structure labelled B.
3. Name the structure labelled C.
4. Name the structure labelled D.
5. Name the structure labelled E.

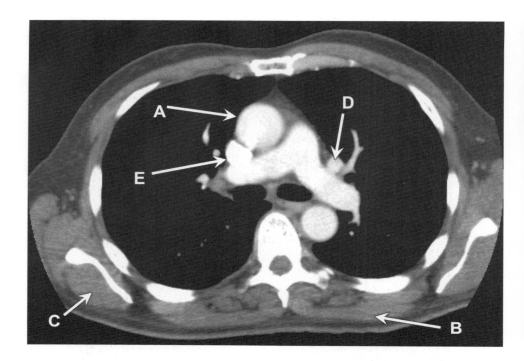

Case 3

1. Name the structure labelled A.
2. Name the structure labelled B.
3. Name the structure labelled C.
4. Name the structure labelled D.
5. Name the structure labelled E.

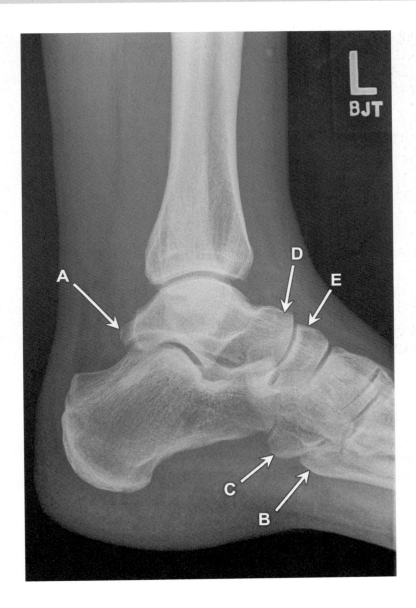

Case 4

1. Name the structure labelled A.

2. Name the structure labelled B.

3. Name the structure labelled C.

4. Name the structure labelled D.

5. Name the structure labelled E.

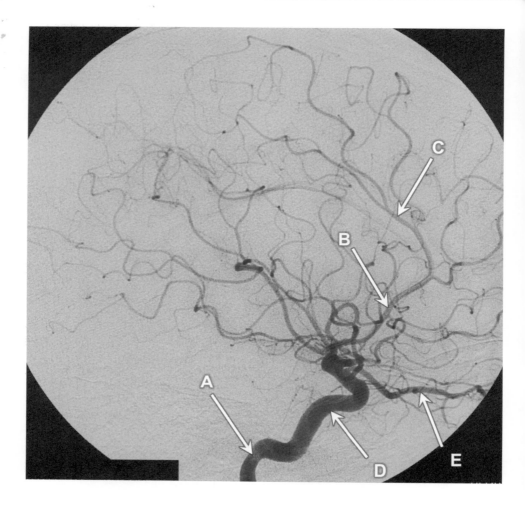

Case 5

1. Name bone that the vessel labelled A is passing through.
2. Name the vessel labelled B.
3. Name the vessel labelled C.
4. Name the vessel labelled D.
5. Name the vessel labelled E.

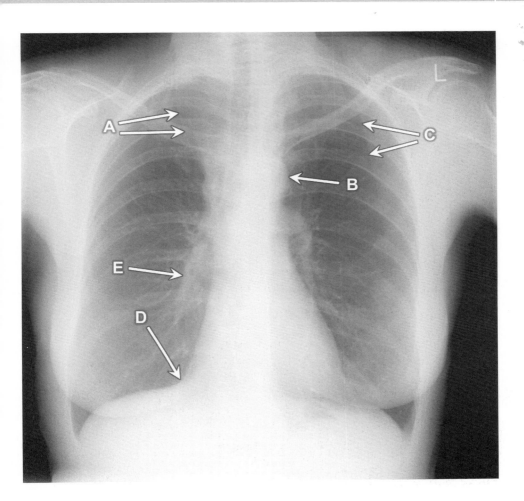

Case 6

1. Name the linear normal variant labelled A.
2. Name the structure labelled B.
3. Name the structure labelled C.
4. Name the angle labelled D.
5. Name the structure labelled E.

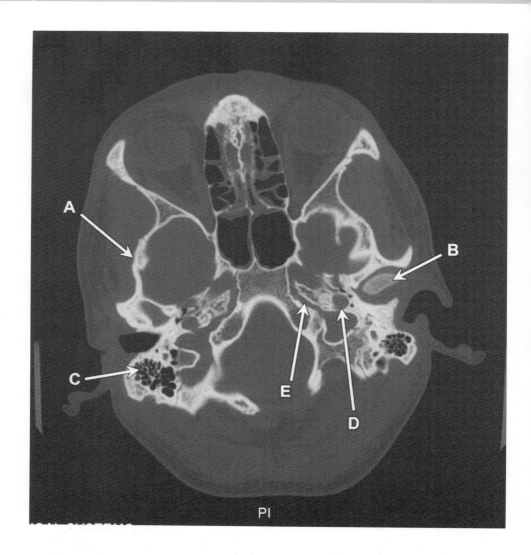

PI

Case 7

1. Name the bone labelled A.
2. Name the structure labelled B.
3. Name the structures labelled C.
4. Name the foramen labelled D.
5. Name the foramen labelled E.

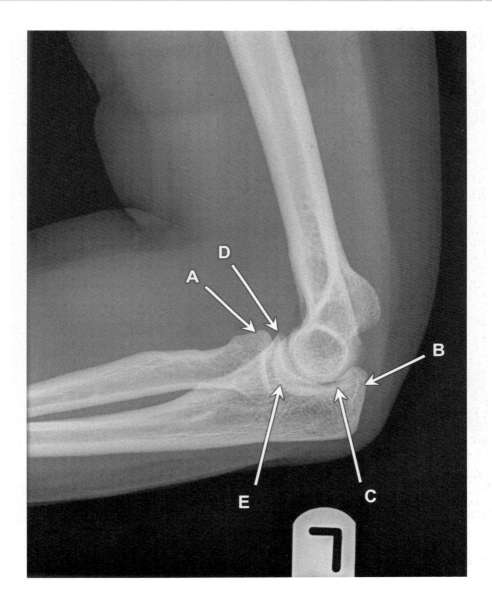

Case 8

1. Name the structure labelled A.
2. Name the structure labelled B.
3. Name the notch labelled C.
4. Name the structure labelled D.
5. Name the structure labelled E.

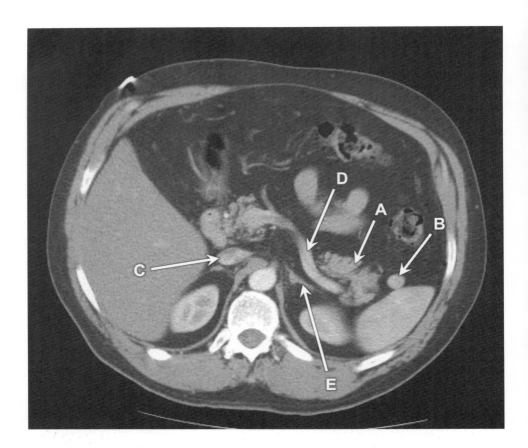

Case 9

1. Name the structure labelled A.

2. Name the structure labelled B.

3. Name the structure labelled C.

4. Name the structure labelled D.

5. Name the structure labelled E.

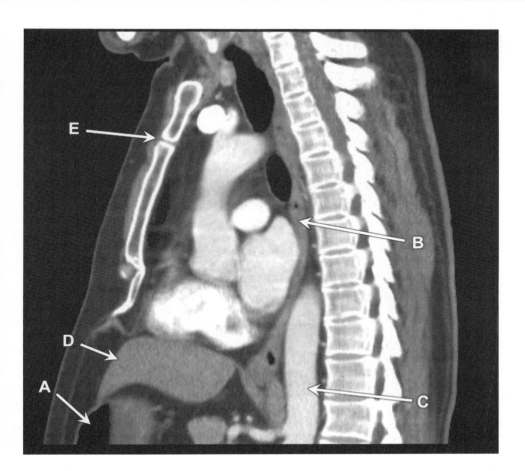

Case 10

1. Name the organ labelled A.
2. Name the structure labelled B.
3. Name the structure labelled C.
4. Name the structure labelled D.
5. Name the structure labelled E.

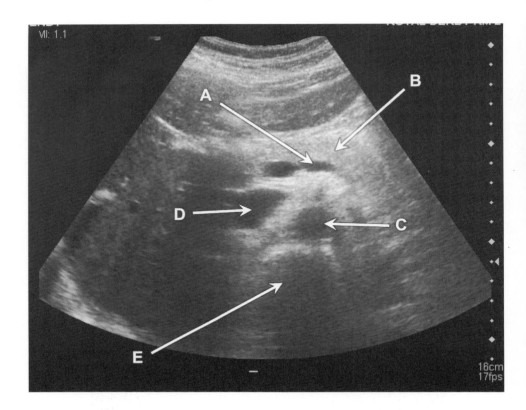

Case 11

1. Name the structure labelled A.
2. Name the structure labelled B.
3. Name the structure labelled C.
4. Name the structure labelled D.
5. Name the structure labelled E.

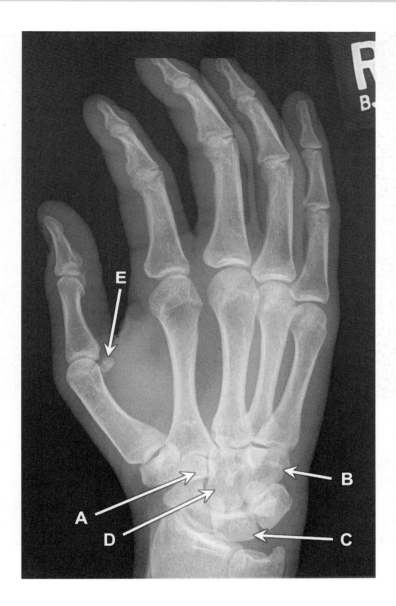

Case 12

1. Name the bone labelled A.

2. Name the bone labelled B.

3. Name the bone labelled C.

4. Name the bone labelled D.

5. Name the structure labelled E.

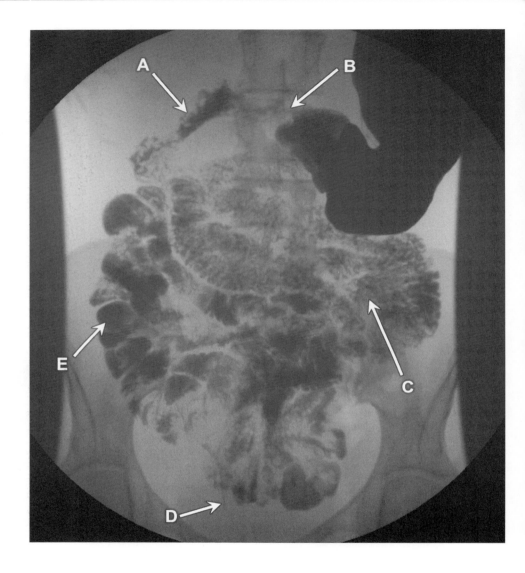

Case 13

1. Name the structure labelled A.

2. Name the structure labelled B.

3. Name the structure labelled C.

4. Name the structure labelled D.

5. Name the structure labelled E.

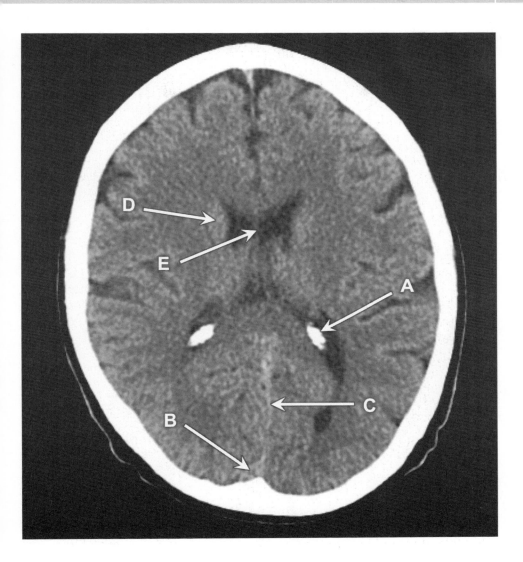

Case 14

1. Name the structure labelled A.
2. Name the structure labelled B.
3. Name the structure labelled C.
4. Name the structure labelled D.
5. Name the structure labelled E.

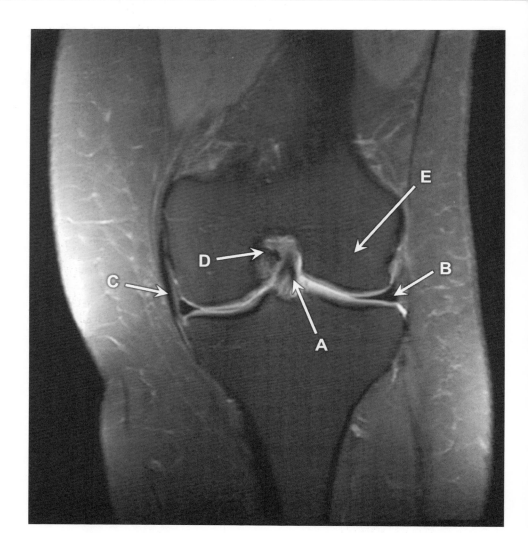

Case 15

1. Name the structure labelled A.
2. Name the structure labelled B.
3. Name the structure labelled C.
4. Name the structure labelled D.
5. Name the structure labelled E.

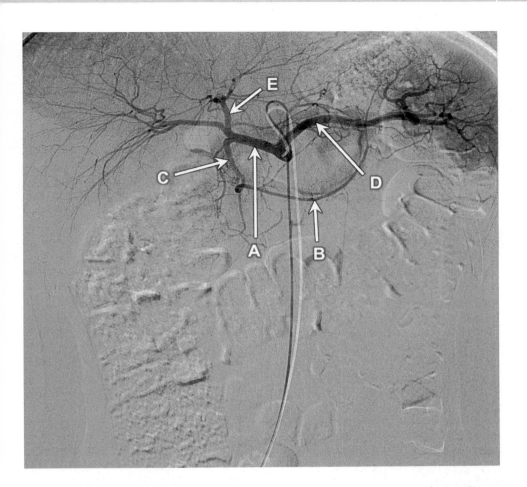

Case 16

1. Name the structure labelled A.
2. Name the structure labelled B.
3. Name the structure labelled C.
4. Name the structure labelled D.
5. Name the structure labelled E.

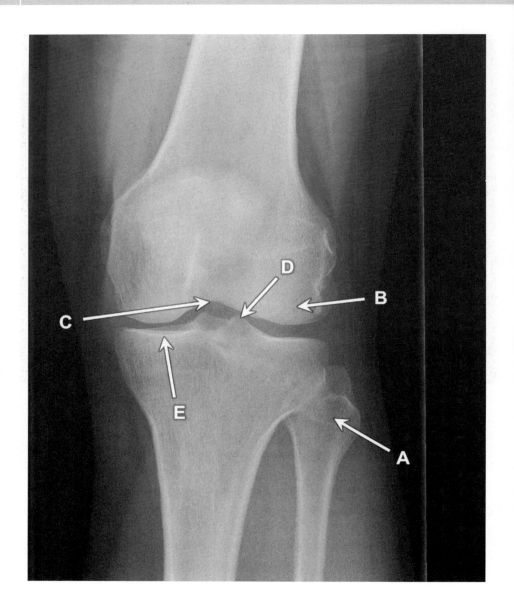

Case 17

1. Name the structure labelled A.
2. Name the structure labelled B.
3. Name the structure labelled C.
4. Name the structure labelled D.
5. Name the structure labelled E.

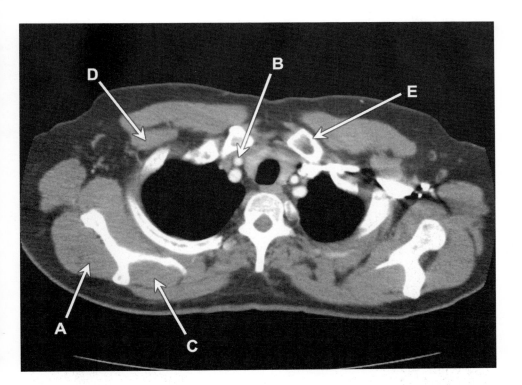

Case 18

1. Name the structure labelled A.
2. Name the structure labelled B.
3. Name the structure labelled C.
4. Name the structure labelled D.
5. Name the structure labelled E.

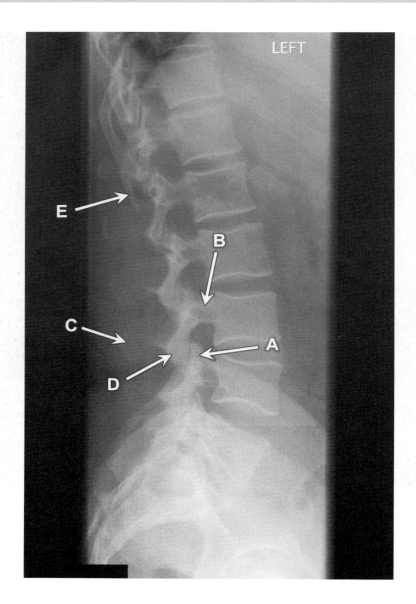

Case 19

1. Name the structure labelled A.

2. Name the structure labelled B.

3. Name the structure labelled C.

4. Name the structure labelled D.

5. Name the structure labelled E.

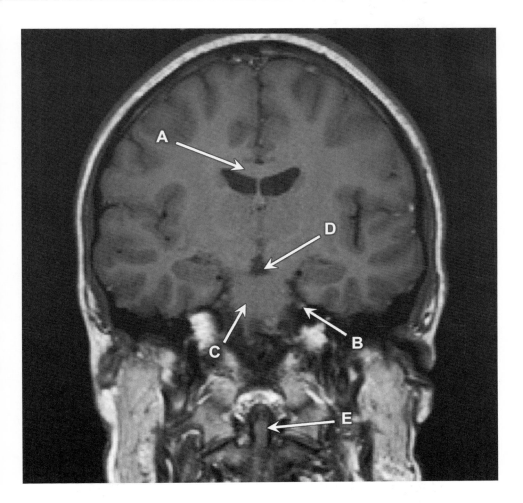

Case 20

1. Name the structure labelled A.
2. Name the structure labelled B.
3. Name the structure labelled C.
4. Name the structure labelled D.
5. Name the structure labelled E.

Case 1

Plain radiograph. AP left shoulder.

1. Left acromion
2. Left greater tuberosity of humerus
3. Left lesser tuberosity of humerus
4. Left acromioclavicular joint
5. Left coracoid process of scapula

Case 2

MRCP

1. Gallbladder
2. Right renal pelvis
3. Left hepatic duct
4. Cystic duct
5. Common bile duct

It is important to remember that structures other than the organs in question are often revealed in these focused examinations. On this image, both renal collecting systems are visible. The other white structure at the bottom right-hand side of the image is the spinal canal.

Case 3

CT thorax. Axial section.

1. Ascending aorta/ aortic root
2. Left trapezius muscle
3. Right infraspinatus muscle
4. Left superior pulmonary vein
5. Superior vena cava

Case 4

Plain radiograph. Lateral left ankle.

1. Left lateral tubercle of the talus
2. Left fifth metatarsal
3. Left cuboid
4. Left head of the talus
5. Left navicular

Case 5

Internal carotid angiogram. Lateral view

1. Petrous temporal bone
2. Anterior cerebral artery
3. Pericallosal artery
4. Internal carotid artery within the cavernous sinus
5. Ophthalmic artery

Case 6

Plain radiograph. AP chest.

1. Azygos fissure
2. Aortic arch
3. Medial border of the left scapula
4. Right cardiophrenic angle
5. Interlobar artery

The azygos fissure is seen in about 1% of the population. The azygos vein sits at the caudal aspect of the fissure.

Case 7

CT base of skull. Axial section.

1. Right squamous temporal bone
2. Left mandibular condyle
3. Right mastoid air cells
4. Left carotid canal
5. Left foramen lacerum

Case 8

Plain radiograph. Lateral elbow.

1. Head of radius
2. Olecranon of ulna
3. Trochlear notch of ulna
4. Coronoid process of ulna
5. Capitulum of humerus

Slightly tricky. Close inspection reveals that the structure labelled E is articulating with the radial head, so must be the capitulum of the humerus. The 'C' arrow, however points to the notch on the ulna.

Case 9

CT abdomen. Axial section.

1. Tail of pancreas
2. Splenunculus
3. Inferior vena cava
4. Splenic vein
5. Left adrenal gland

Splenunculi are small nodules of spleen, detached from the rest of the organ. They are seen in between 15 and 30% of people and are of no clinical significance.

Case 10

CT thorax. Sagittal section.

1. Stomach
2. Oesophagus
3. Descending aorta
4. Left lobe of liver
5. Manubriosternal joint (sternal angle)

Case 11

Ultrasound abdomen. Transverse section at the epigastrium.

1. Splenic vein
2. Body of pancreas
3. Abdominal aorta
4. Inferior vena cava
5. Vertebral body

Case 12

Plain radiograph. Oblique hand and wrist.

1. Right trapezoid
2. Right hamate
3. Right lunate
4. Right capitate
5. Sesamoid bone

Case 13

Barium follow-through. 'Overcouch' AP radiograph.

1. Duodenum (second part)
2. Pylorus
3. Jejunum
4. Appendix
5. Ascending colon

Case 14

CT brain. Axial section

1. Left choroid plexus (calcified)
2. Superior sagittal sinus
3. Falx cerebri
4. Head of the right caudate nucleus
5. Septum pellucidum

Case 15

MRI knee. T1W coronal section.

1. Anterior cruciate ligament
2. Lateral meniscus
3. Medial collateral ligament
4. Posterior cruciate ligament
5. Lateral femoral condyle

Case 16

Coeliac axis angiogram.

1. Common hepatic artery
2. Right gastroepiploic artery
3. Gastroduodenal artery
4. Splenic artery
5. Left hepatic artery

Several normal variants may exist in the arterial anatomy here, particularly in relation to the origin of the hepatic arteries. Candidates should be aware of these. Choose your favourite anatomy book and have a look.

Case 17

Plain radiograph. AP knee.

1. Head of the fibula
2. Lateral femoral condyle
3. Intercondylar fossa
4. Lateral tibial spine
5. Medial tibial plateau

Case 18

CT chest. Axial section.

1. Right infraspinatus muscle
2. Right common carotid artery
3. Right supraspinatus muscle
4. Right pectoralis minor muscle
5. Head of left clavicle

Case 19

Plain radiograph. Lateral lumbar spine.

1. Superior articular process of L4
2. Pedicle of L3
3. Spinous process of L3
4. Inferior articular process of L3
5. Twelfth rib

Case 20

MRI brain. T1W coronal section.

1. Corpus callosum
2. Left trigeminal nerve
3. Pons
4. Interpeduncular cistern
5. Odontoid process (peg) of C2

You have 75 minutes to complete the examination

Difficulty rating: Ben Nevis

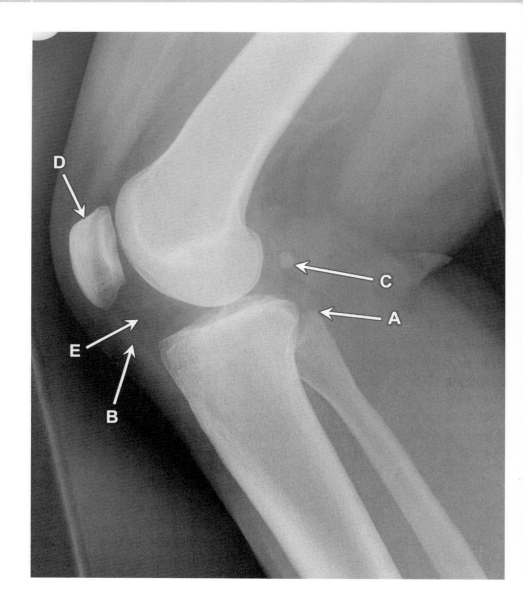

Case 1

1. Name the structure labelled A.
2. Name the structure labelled B.
3. Name the structure labelled C.
4. Name the structure labelled D.
5. Name the structure that fills that space labelled E.

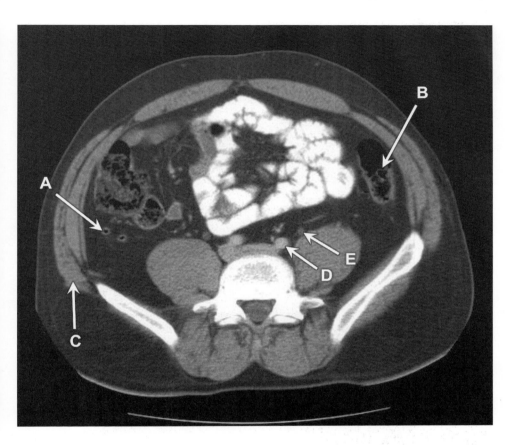

Case 2

1. Name the structure labelled A.

2. Name the structure labelled B.

3. Name the structure labelled C.

4. Name the structure labelled D.

5. Name the structure labelled E.

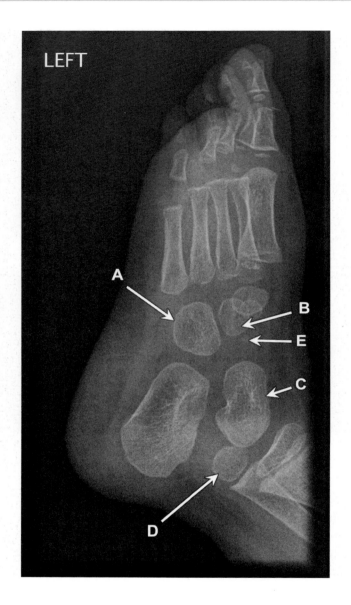

Case 3

1. Name the structure labelled A.

2. Name the structure labelled B.

3. Name the structure labelled C.

4. Name the structure labelled D.

5. Name the structure labelled E.

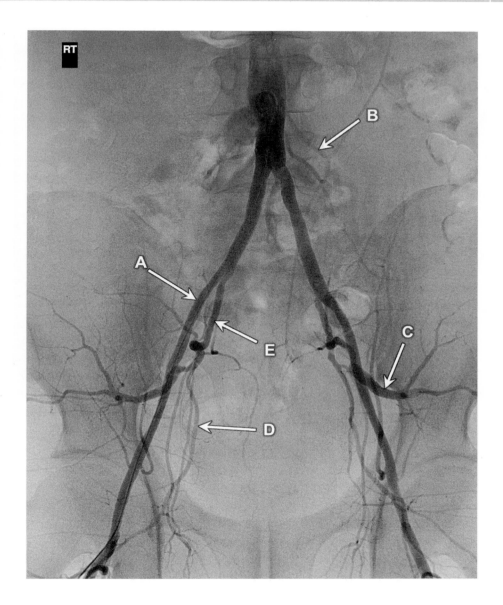

Case 4

1. Name the vessel labelled A.

2. Name the vessel labelled B.

3. Name the vessel labelled C.

4. Name the vessel labelled D.

5. Name the vessel labelled E.

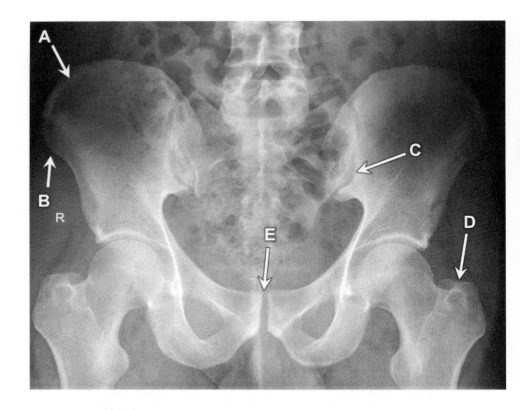

Case 5

1. Name the structure labelled A.
2. Name the structure labelled B.
3. Name the structure labelled C.
4. Name the structure labelled D.
5. Name the structure labelled E.

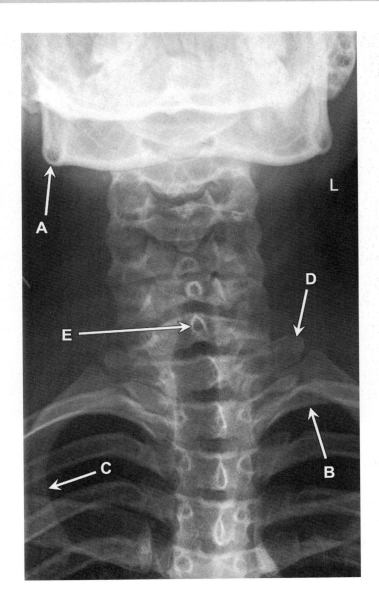

Case 6

1. Name the structure labelled A.

2. Name the structure labelled B.

3. Name the structure labelled C.

4. Name the structure labelled D.

5. Name the structure labelled E.

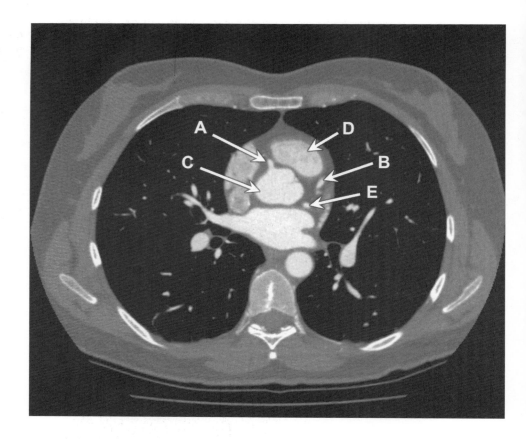

Case 7

1. Name the structure labelled A.
2. Name the structure labelled B.
3. Name the structure labelled C.
4. Name the structure labelled D.
5. Name the structure labelled E.

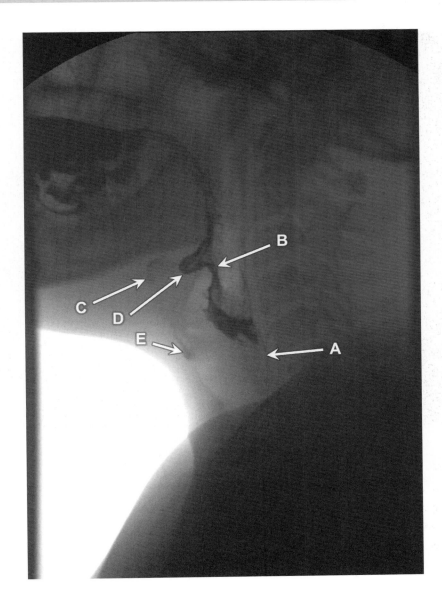

Case 8

1. Name the structure labelled A.

2. Name the structure labelled B.

3. Name the structure labelled C.

4. Name the structure labelled D.

5. Name the structure labelled E.

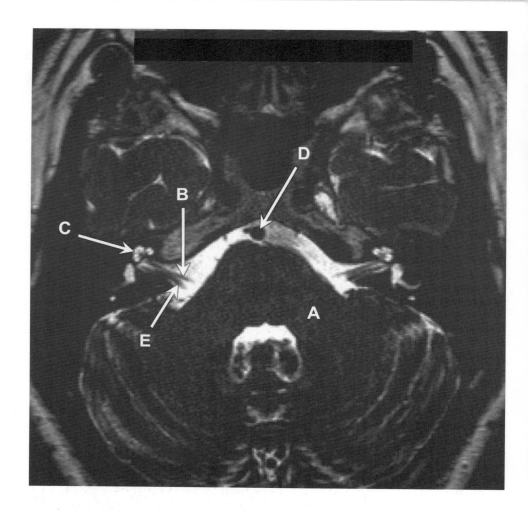

Case 9

1. Name the structure labelled A.
2. Name the structure labelled B.
3. Name the structure labelled C.
4. Name the structure labelled D.
5. Name the structure labelled E.

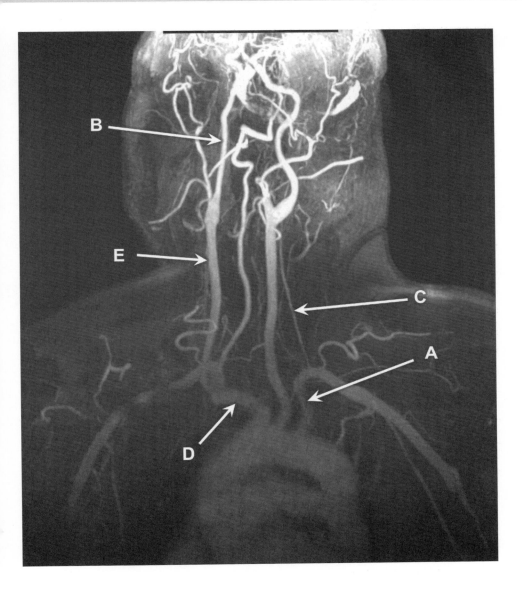

Case 10

1. Name the structure labelled A.

2. Name the structure labelled B.

3. Name the structure labelled C.

4. Name the structure labelled D.

5. Name the structure labelled E.

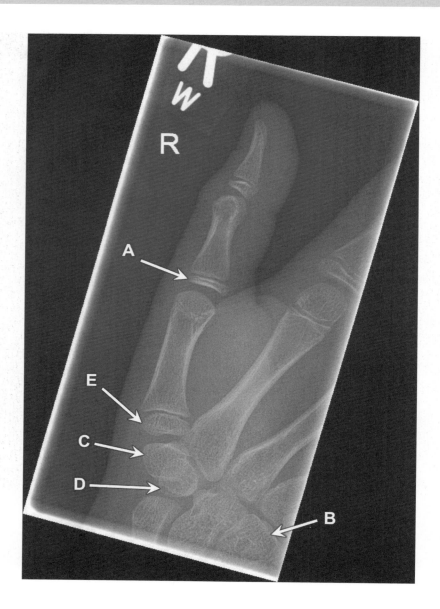

Case 11

1. Name the structure labelled A.
2. Name the structure labelled B.
3. Name the structure labelled C.
4. Name the structure labelled D.
5. Name the structure labelled E.

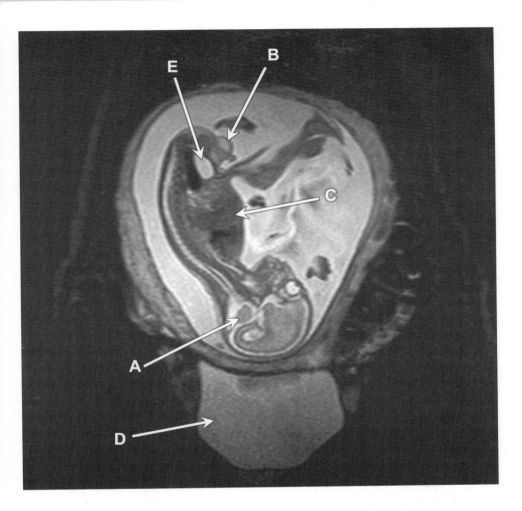

Case 12

1. Name the structure labelled A.
2. Name the structure labelled B.
3. Name the structure labelled C.
4. Name the structure labelled D.
5. Name the structure labelled E.

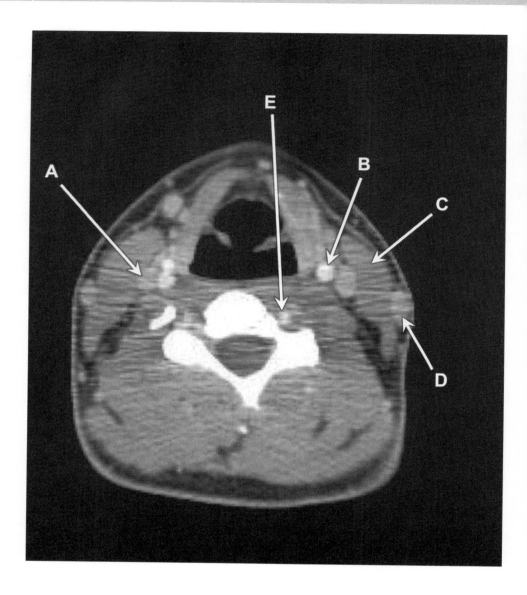

Case 13

1. Name the structure labelled A.
2. Name the structure labelled B.
3. Name the structure labelled C.
4. Name the structure labelled D.
5. Name the structure labelled E.

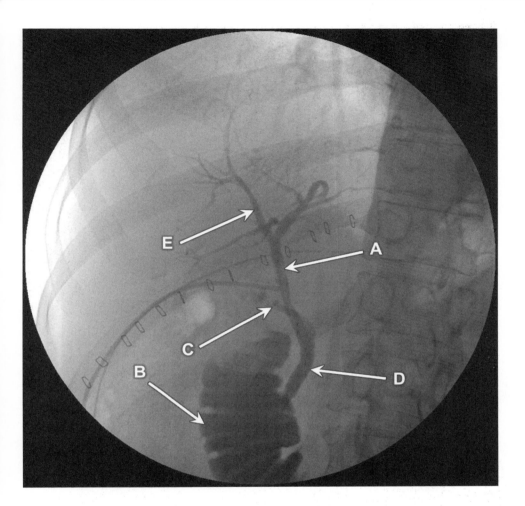

Case 14

1. Name the structure labelled A.
2. Name the structure labelled B.
3. Name the structure labelled C.
4. Name the structure labelled D.
5. Name the structure labelled E.

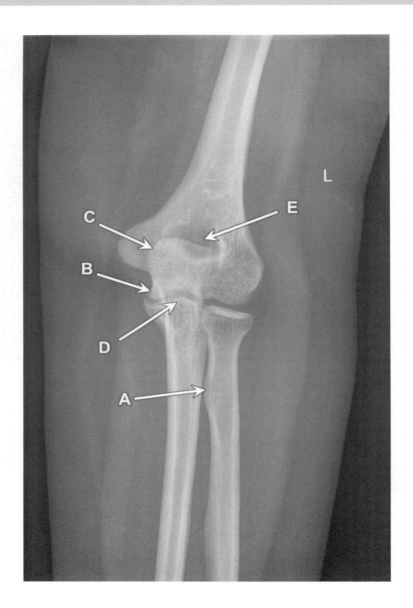

Case 15

1. Name the structure labelled A.
2. Name the structure labelled B.
3. Name the structure labelled C.
4. Name the structure labelled D.
5. Name the structure labelled E.

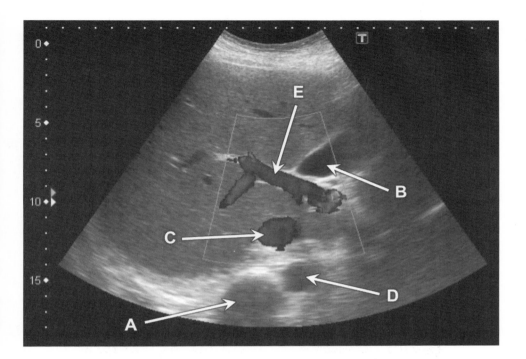

Case 16

1. Name the structure labelled A.
2. Name the structure labelled B.
3. Name the structure labelled C.
4. Name the structure labelled D.
5. Name the structure labelled E.

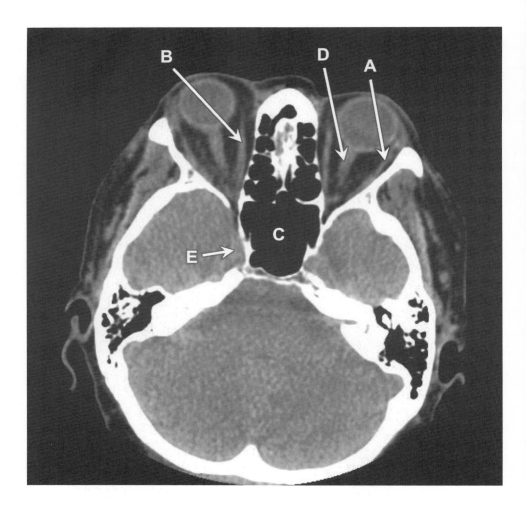

Case 17

1. Name the structure labelled A.

2. Name the structure labelled B.

3. Name the structure labelled C.

4. Name the structure labelled D.

5. Name the vessel that passes through the area labelled E.

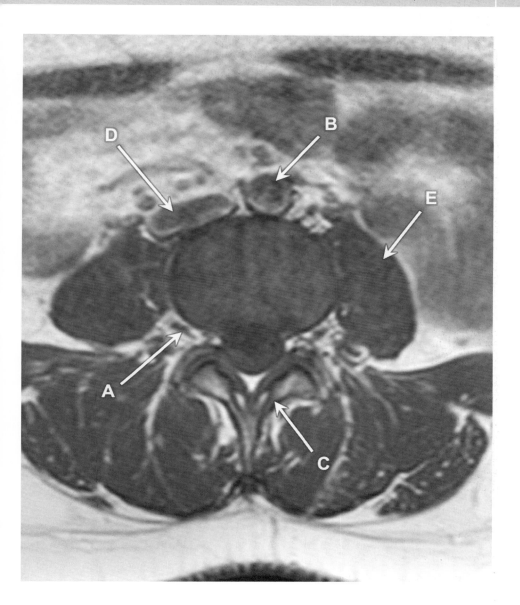

Case 18

1. Name the structure labelled A.
2. Name the structure labelled B.
3. Name the structure labelled C.
4. Name the structure labelled D.
5. Name the structure labelled E.

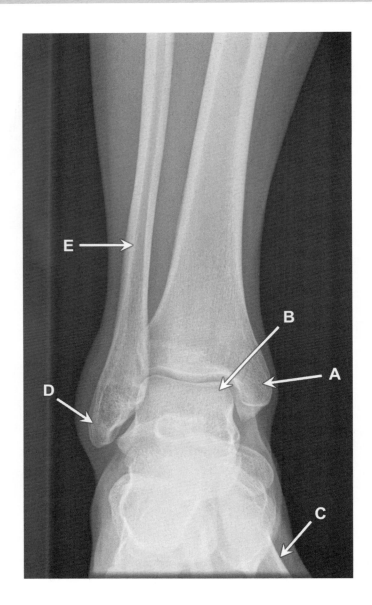

Case 19

1. Name the structure labelled A.
2. Name the structure labelled B.
3. Name the structure labelled C.
4. Name the structure labelled D.
5. Name the structure labelled E.

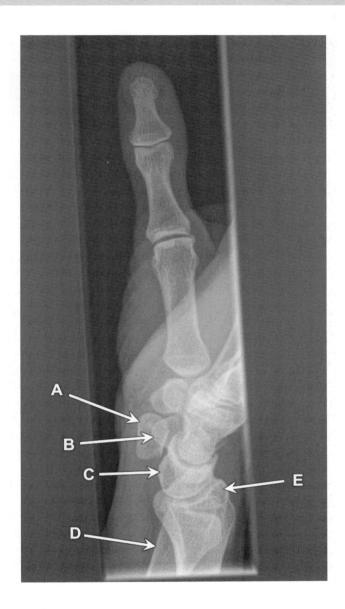

Case 20

1. Name the structure labelled A.
2. Name the structure labelled B.
3. Name the structure labelled C.
4. Name the structure labelled D.
5. Name the structure labelled E.

Case 1

Lateral radiograph, knee.

1. Styloid process of the fibula
2. Patellar tendon
3. Fabella
4. Patella
5. Hoffa's fat pad

A fabella is a small accessory bone that lies within the lateral head of the gastrocnemius muscle.

Case 2

CT abdomen with iv and oral contrast. Axial section.

1. Appendix
2. Descending colon
3. Right external oblique
4. Left common iliac artery
5. Left ureter

Case 3

Paediatric foot. DP oblique.

1. Left cuboid
2. Left lateral cuneiform
3. Left talus
4. Left distal fibular epiphysis
5. Left ossification centre of the navicula

Case 4

Aortogram.

1. Right external iliac artery
2. Left colic artery
3. Left superior gluteal artery
4. Right inferior gluteal artery
5. Right internal iliac artery

Case 5

AP radiograph. Pelvis.

1. Right iliac crest
2. Right anterior superior iliac spine
3. Left sacroiliac joint
4. Left greater trochanter of the femur
5. Symphysis pubis

Case 6

AP radiograph. Cervical spine.

1. Right angle of the mandible
2. Left second rib
3. Right first rib
4. Left transverse process of T1
5. Spinous process of C7

Case 7

CT coronary angiogram. Axial section.

1. Right main coronary artery
2. Left anterior descending artery
3. Aortic root
4. Right ventricular outflow tract
5. Circumflex artery

Case 8

Barium swallow. Lateral view.

1. Oesophagus
2. Epiglottis
3. Hyoid bone
4. Vallecula
5. Thyroid cartilage

Case 9

MRI internal auditory meati. T2 axial section

1. Left cerebellar peduncle
2. Right facial (VII) nerve
3. Right cochlea
4. Basilar artery
5. Right vestibulocochlear (VIII) nerve

Case 10

MR-angiogram. Aortic arch/ neck (MIP). Oblique view.

1. Left subclavian artery
2. Right internal carotid artery
3. Left vertebral artery
4. Right brachiocephalic (innominate) artery
5. Right common carotid artery

Case 11

AP radiograph. Right thumb (paediatric)

1. Right proximal epiphysis of the proximal phalanx of the thumb
2. Right hamate
3. Right trapezium
4. Right trapezoid
5. Right proximal epiphysis of the first metacarpal

It is potentially easy to mistake unfused epiphyses for separate bones, for example the structure labelled E. Beware.

Case 12

Fetal MRI. T2-weighted coronal section.

1. Fetal cerebellum
2. Fetal scrotum
3. Fetal liver
4. Maternal urinary bladder
5. Fetal urinary bladder

Case 13

CT neck. Axial section.

1. Right internal jugular vein
2. Left common carotid artery
3. Sternocleidomastoid
4. Left external jugular vein
5. Left vertebral artery

Case 14

T-tube cholangiogram (post cholecystectomy).

1. Common hepatic duct
2. Duodenum
3. Cystic duct remnant
4. Common bile duct
5. Right hepatic duct

The line extending from the common hepatic duct to the left of the image is the tube leading to the external drain. Surgical staples can be seen projected over the image giving further clue that this is a postoperative study.

Case 15

Plain radiograph. AP elbow.

1. Radial tuberosity (left)
2. Trochlea (left)
3. Olecranon process of the ulna (left)
4. Coronoid process of the ulna (left)
5. Olecranon fossa (left)

Case 16

Abdominal ultrasound. Oblique view at the porta hepatis.

1. Spinal column
2. Gallbladder
3. Inferior vena cava
4. Abdominal aorta
5. Hepatic portal vein

Case 17

CT orbits, non-contrast. Axial section.

1. Left lateral rectus
2. Right medial rectus
3. Sphenoid sinus
4. Left optic nerve
5. Right internal carotid artery

Case 18

MRI lumbar spine. T1W axial section.

1. Exiting nerve root (right)
2. Abdominal aorta
3. Left vertebral lamina
4. Inferior vena cava
5. Left psoas muscle

Case 19

AP radiograph of the ankle

1. Medial malleolus (of the tibia)
2. Talus
3. First metatarsal
4. Lateral malleolus (of the fibula)
5. Fibula

Case 20

Lateral radiograph of the thumb

1. Pisiform
2. Scaphoid tubercle
3. Lunate
4. Ulna
5. Radius

Differentiating the ulna and radius is straightforward with a little careful examination. The bone labelled E is seen to articulate with the lunate so must be the radius; the bone labelled D does not and has the characteristic styloid process.

You have 75 minutes to complete the examination

Difficulty rating: Ben Nevis

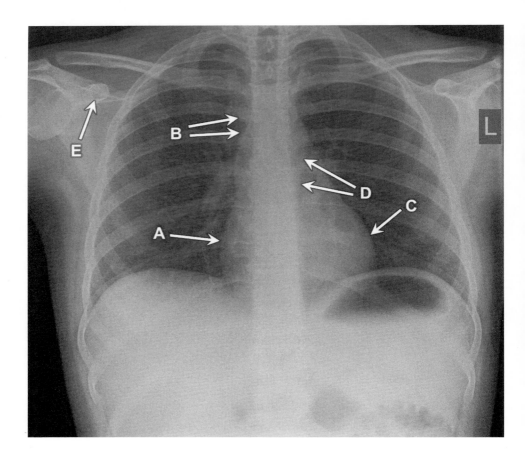

Case 1

1. Name the structure labelled A.
2. Name the structure forming the mediastinal border labelled B.
3. Name the structure labelled C.
4. Name the structure labelled D.
5. Name the muscle that connects the structure labelled E and the thoracic cage.

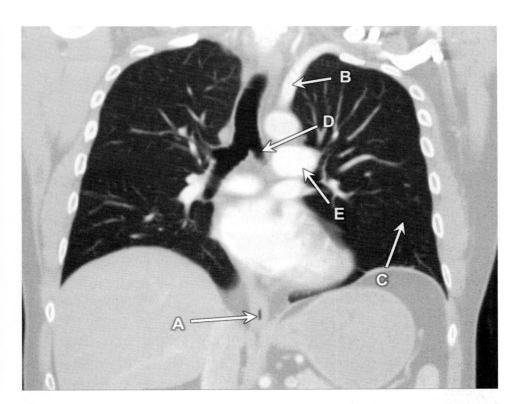

Case 2

1. Name the structure labelled A.
2. Name the structure labelled B.
3. Name the linear structure labelled C.
4. Name the structure labelled D.
5. Name the structure labelled E.

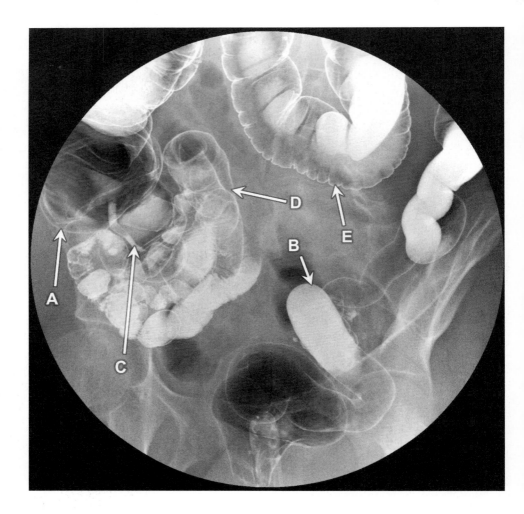

Case 3

1. Name the structure labelled A.
2. Name the structure labelled B.
3. Name the structure labelled C.
4. Name the structure labelled D.
5. Name the structure labelled E.

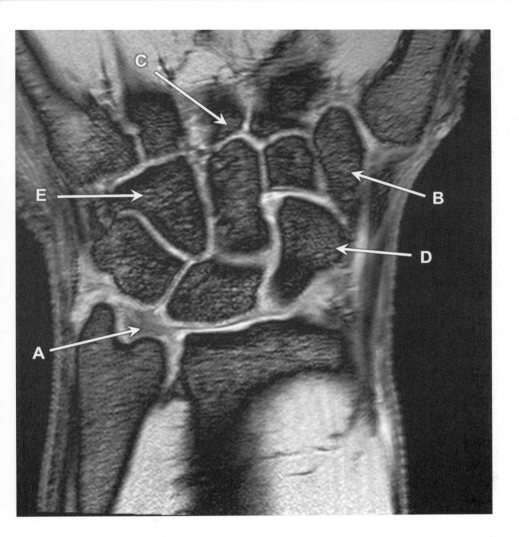

Case 4

1. Name the structure labelled A.
2. Name the structure labelled B.
3. Name the structure labelled C.
4. Name the structure labelled D.
5. Name the structure labelled E.

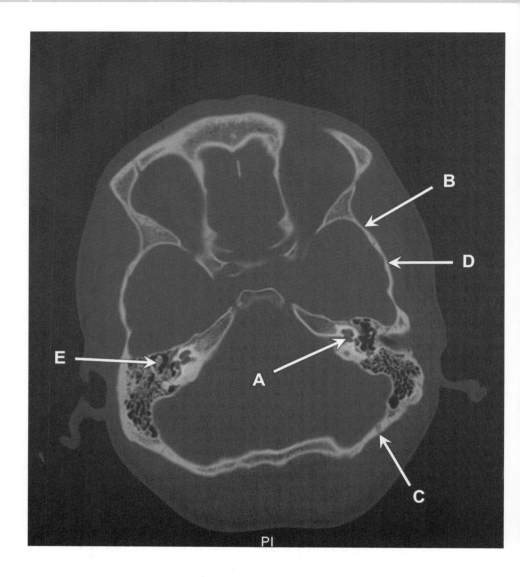

Case 5

1. Name the structure labelled A.
2. Name the structure labelled B.
3. Name the structure labelled C.
4. Name the structure labelled D.
5. Name the structure labelled E.

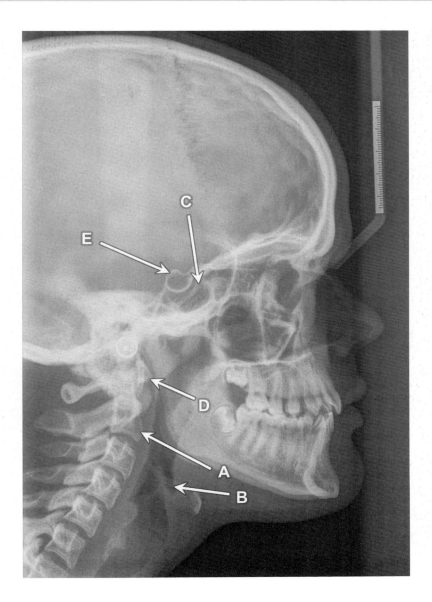

Case 6

1. Name the structure labelled A.
2. Name the space labelled B.
3. Name the structure labelled C.
4. Name the structure labelled D.
5. Name the structure labelled E.

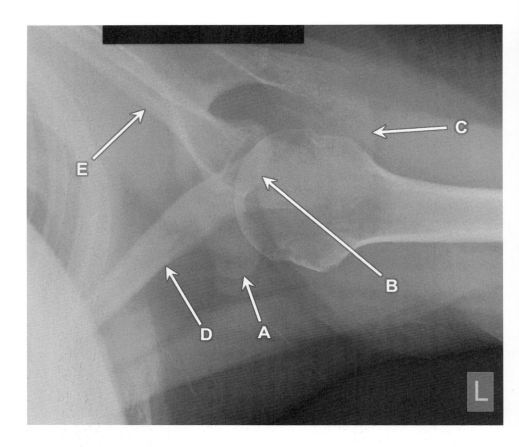

Case 7

1. Name the structure labelled A.
2. Name the structure labelled B.
3. Name the structure labelled C.
4. Name the structure labelled D.
5. Name the structure labelled E.

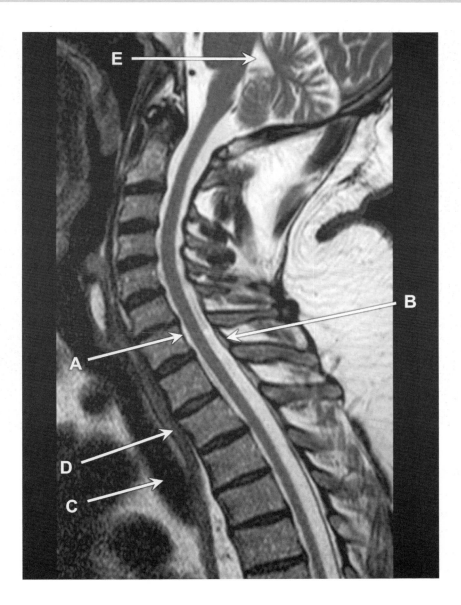

Case 8

1. Name the structure labelled A.
2. Name the structure labelled B.
3. Name the structure labelled C.
4. Name the structure labelled D.
5. Name the structure labelled E.

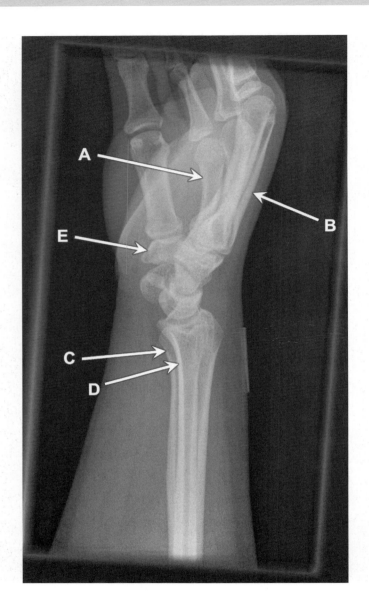

Case 9

1. Name the structure labelled A.
2. Name the structure labelled B.
3. Name the structure labelled C.
4. Name the structure labelled D.
5. Name the structure labelled E.

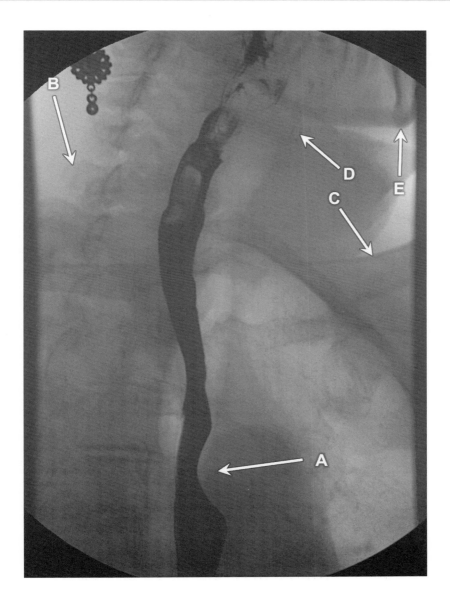

Case 10

1. Name the structure that causes the oesophageal indentation labelled A.

2. Name the structure labelled B.

3. Name the structure labelled C.

4. Name the structure labelled D.

5. Name the structure labelled E.

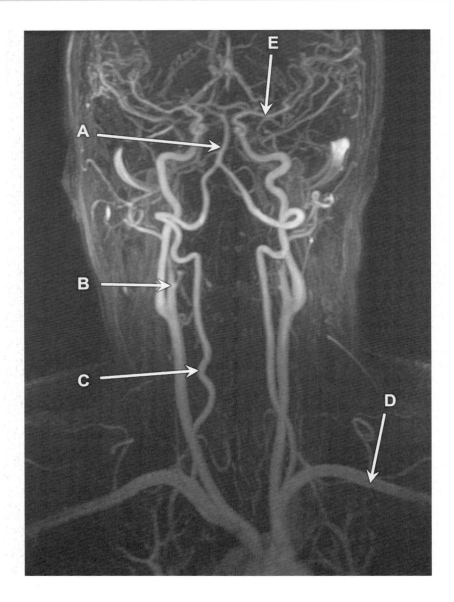

Case 11

1. Name the structure labelled A.

2. Name the structure labelled B.

3. Name the structure labelled C.

4. Name the structure labelled D.

5. Name the structure labelled E.

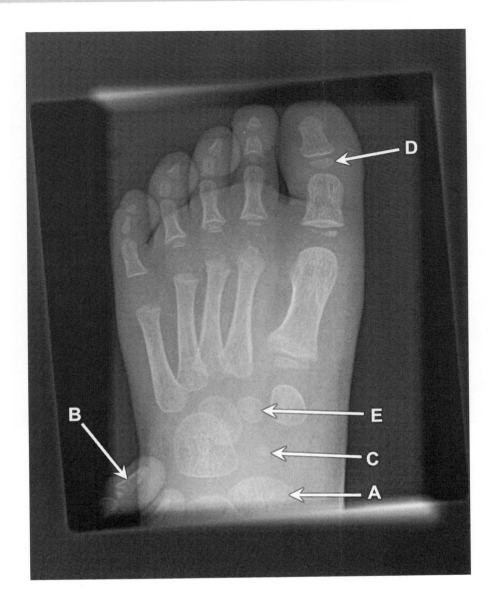

Case 12

1. Name the structure labelled A.
2. Name the structure labelled B.
3. Name the structure labelled C.
4. Name the structure labelled D.
5. Name the structure labelled E.

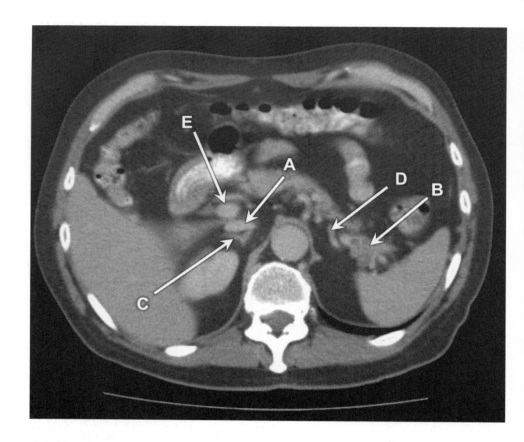

Case 13

1. Name the structure labelled A.

2. Name the structure labelled B.

3. Name the structure labelled C.

4. Name the structure labelled D.

5. Name the structure labelled E.

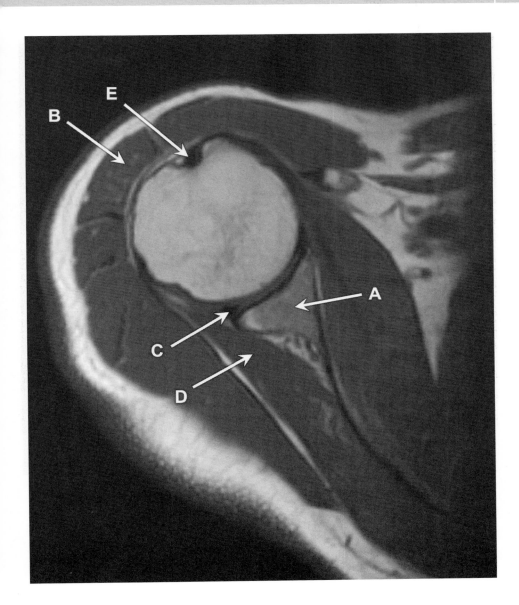

Case 14

1. Name the structure labelled A.
2. Name the structure labelled B.
3. Name the structure labelled C.
4. Name the structure labelled D.
5. Name the structure labelled E.

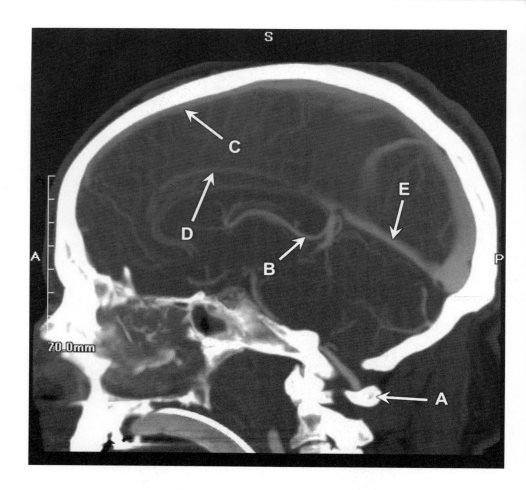

Case 15

1. Name the structure labelled A.
2. Name the structure labelled B.
3. Name the structure labelled C.
4. Name the structure labelled D.
5. Name the structure labelled E.

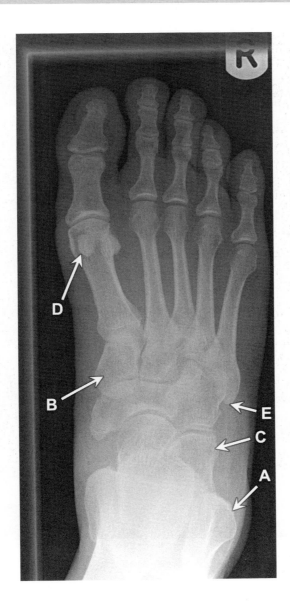

Case 16

1. Name the structure labelled A.
2. Name the structure labelled B.
3. Name the structure labelled C.
4. Name the structure labelled D.
5. Name the tendon inserting on the structure labelled E.

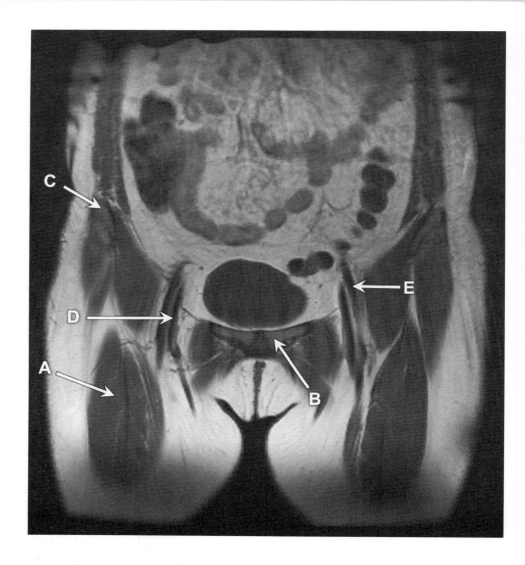

Case 17

1. Name the structure labelled A.
2. Name the structure labelled B.
3. Name the structure labelled C.
4. Name the structure labelled D.
5. Name the structure labelled E.

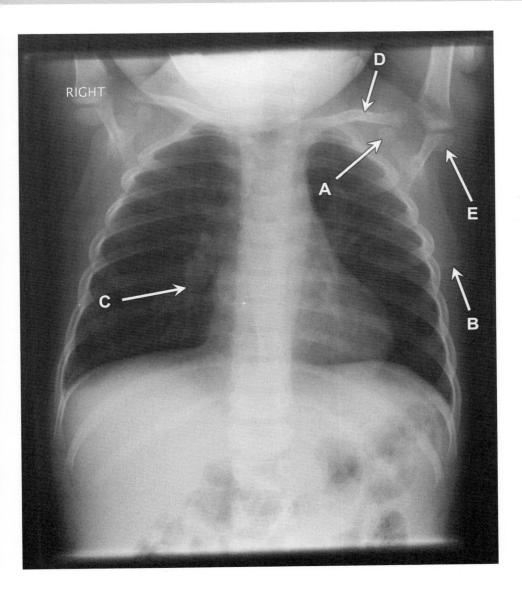

Case 18

1. Name the structure labelled A.

2. Name the structure labelled B.

3. Name the structure labelled C.

4. Name the structure labelled D.

5. Name the structure labelled E.

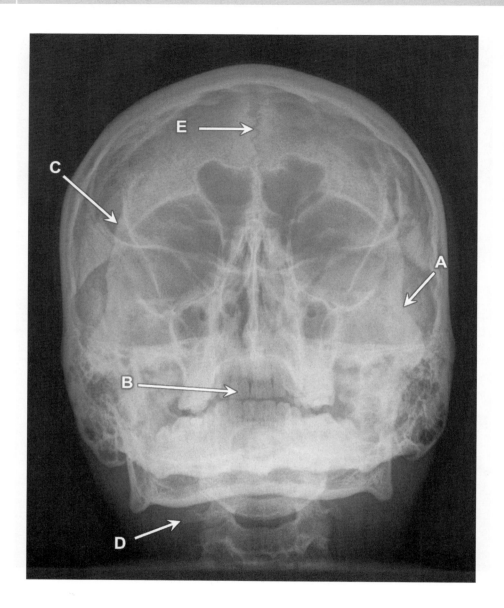

Case 19

1. Name the structure labelled A.
2. Name the structure labelled B.
3. Name the structure labelled C.
4. Name the structure labelled D.
5. Name the structure labelled E.

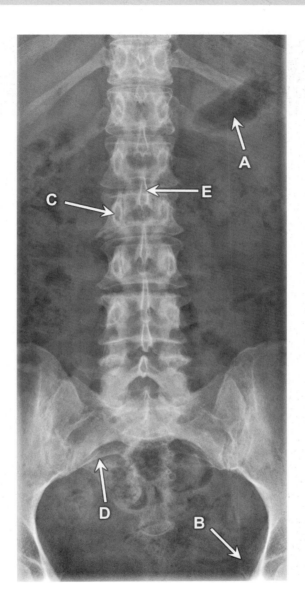

Case 20

1. Name the structure labelled A.
2. Name the structure labelled B.
3. Name the structure labelled C.
4. Name the structure labelled D.
5. Name the structure labelled E.

Case 1

PA chest radiograph.

1. Right atrium
2. Superior vena cava
3. Left ventricle
4. Descending thoracic aorta
5. Right pectoralis minor

Case 2

CT Chest, coronal section.

1. Oesophagus
2. Left subclavian artery
3. Left oblique fissure
4. Left main bronchus
5. Left main pulmonary artery

Case 3

Oblique pelvic 'spot' image from a double-contrast barium enema.

1. Caecum
2. Sigmoid colon
3. Appendix
4. Ileum
5. Transverse colon

Case 4

MR wrist, coronal section, T1 weighted.

1. Triangular fibrocartilage
2. Trapezium
3. Base of third metacarpal bone
4. Scaphoid
5. Hamate

Case 5

CT base of skull, axial image.

1. Left cochlea
2. Left zygomatic bone
3. Left lambdoid suture
4. Squamous part of the left temporal bone
5. Right incus

As well as the bones of the skull, it is important to know the bones of the middle ear and to have an appreciation of those parts of the inner ear that can be seen on CT.

Case 6

Cephalogram

1. Body of the C2 vertebra
2. Vallecula
3. Sphenoid sinus
4. Body of the C1 vertebra (anterior arch)
5. Posterior clinoid process of the sphenoid bone

Case 7

Axial view of the shoulder.

1. Left coracoid process
2. Left glenoid
3. Left acromium
4. Left clavicle
5. Left scapula blade

In this view, it may be easy to mistake the scapular blade for the scapular spine. The spine, however, lies more posterior than this and is continuous with the acromium.

Case 8

MR C-spine, T2-weighted, sagittal.

1. Posterior longitudinal ligament
2. Ligamentum flavum
3. Trachea
4. Oesophagus
5. Fourth ventricle

Case 9

Lateral wrist radiograph.

1. Fifth metacarpal
2. Second metacarpal
3. Ulna
4. Radius
5. Trapezium

This is tricky but is achievable by looking at the articulations. 'A' is the smallest metacarpal so almost certainly number 5. 'B' is the second longest but too far away from 'A' to be the fourth. 'D' articulates directly with the lunate, so is the radius; C does not. 'E' articulates with the first metacarpal so is the trapezium.

Case 10

Barium swallow. Oblique 'spot' view of the upper oesophagus.

1. Aortic arch
2. Spinous process of C7 vertebra
3. Left clavicle
4. Thyroid cartilage
5. Mental process of the mandible

Why the left clavicle? The patient is rotated to their left and an oblique view has been taken through their upper chest. The mediastinal structures are projected left of centre. If this were a PA view in the right lateral oblique position, the aortic arch would be on the other side of the image.

Case 11

MR angiogram. Neck vessels. AP view.

1. Basilar artery
2. Right external carotid artery
3. Right vertebral artery
4. Left subclavian artery
5. Left middle cerebral artery

The carotids can be confused because the 'external' carotids appear to branch medially. Follow the outer branch and you see that it becomes the middle cerebral artery so is the internal carotid artery.

Case 12

DP radiograph of a child's foot.

1. Calcaneum
2. Distal phalanx of a parent's finger
3. Ossification centre of the navicula
4. Proximal epiphysis of the distal phalanx of the great toe
5. Middle cuneiform

Case 13

CT abdomen. Axial section

1. Inferior vena cava
2. Tail of the pancreas
3. Right adrenal gland
4. Splenic artery
5. Portal vein

Case 14

MRI shoulder. T1 weighted. Axial.

1. Glenoid
2. Deltoid muscle
3. Glenoid labrum
4. Infrapinatus muscle
5. Long head of biceps tendon

Case 15

CT cerebral venogram. Sagittal MIP.

1. Posterior arch of C1 vertebra
2. Great cerebral vein (of Galen)
3. Superior sagittal sinus
4. Inferior sagittal sinus
5. Straight sinus

Case 16

DP radiograph of the right foot.

1. Right fibular head/ lateral malleolus
2. Right medial cuneiform
3. Right calcaneum
4. Right tibial sesamoid bone
5. Right peroneus brevis

Case 17

MRI pelvis. T1 weighted. Coronal section

1. Right rectus femoris muscle
2. Left pubic tubercle
3. Right anterior superior iliac spine
4. Right common femoral vein
5. Left common femoral artery

Case 18

PA chest radiograph, child.

1. Ossification centre of the coracoid process of the left scapula
2. Tip of the left scapula
3. Right lower lobe pulmonary artery
4. Left clavicle
5. Proximal epiphysis of the left humerus

Case 19

AP skull radiograph.

1. Left zygoma
2. Right first upper incisor
3. Zygomatic process of the right frontal bone
4. Right lateral process of the C2 vertebra
5. Sagittal suture

Case 20

AP radiograph. Lumbar spine.

1. Stomach
2. Left ischial spine
3. Right pedicle of L2 vertebra
4. Right sacral ala
5. Spinous process of L1 vertebra

Remember that the spinous processes are angled caudally so usually are projected over the vertebral body below.

You have 75 minutes to complete the examination

Difficulty rating: Kilimanjaro

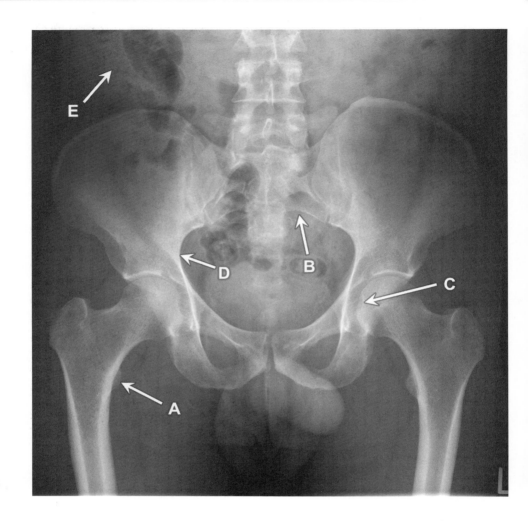

Case 1

1. Name the structure labelled A.
2. Name the structure labelled B.
3. Name the structure labelled C.
4. Name the structure labelled D.
5. Name the structure labelled E.

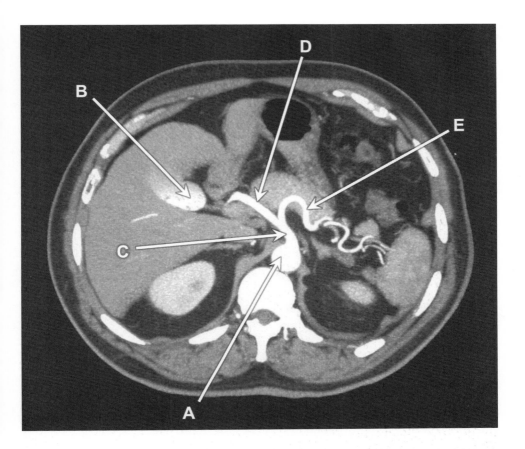

Case 2

1. Name the structure labelled A.

2. Name the structure labelled B.

3. Name the structure labelled C.

4. Name the structure labelled D.

5. Name the structure labelled E.

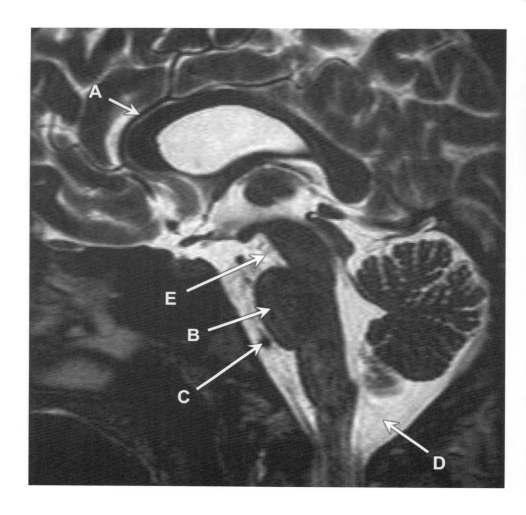

Case 3

1. Name the structure labelled A.
2. Name the structure labelled B.
3. Name the structure labelled C.
4. Name the structure labelled D.
5. Name the structure labelled E.

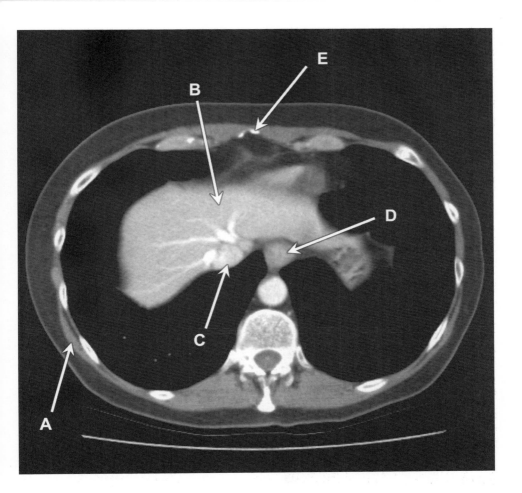

Case 4

1. Name the structure labelled A.
2. Name the segment of the liver labelled B.
3. Name the structure labelled C.
4. Name the structure labelled D.
5. Name the structure labelled E.

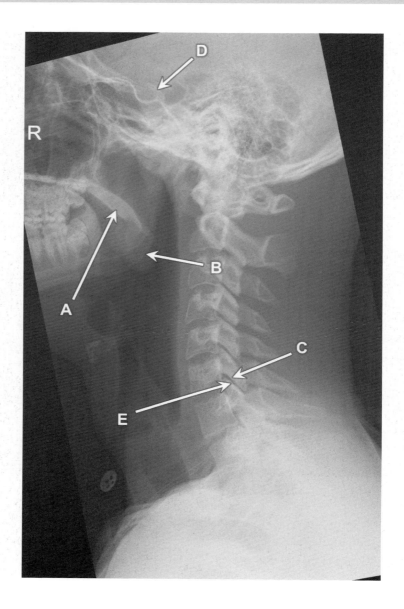

Case 5

1. Name the structure labelled A.

2. Name the bony structure labelled B.

3. Name the structure labelled C.

4. Name the structure labelled D.

5. Name the structure labelled E.

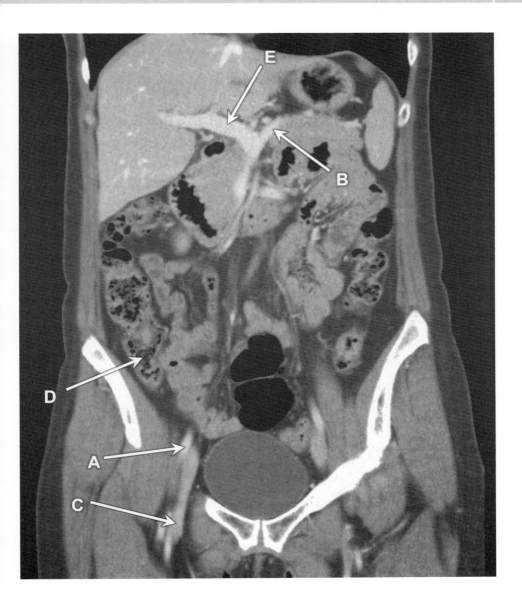

Case 6

1. Name the structure labelled A.
2. Name the structure labelled B.
3. Name the structure labelled C.
4. Name the structure labelled D.
5. Name the structure labelled E.

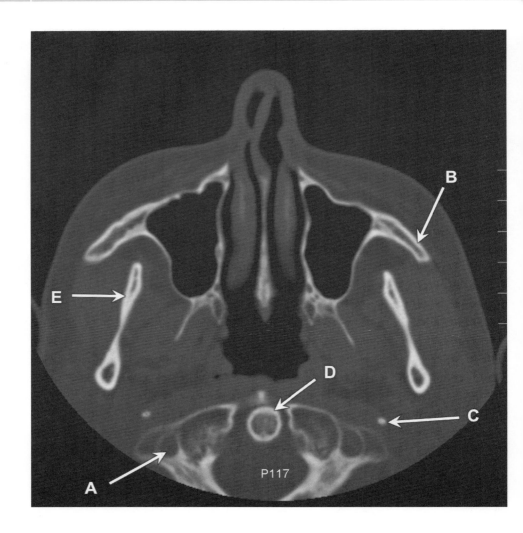

Case 7

1. Name the structure labelled A.
2. Name the structure labelled B.
3. Name the structure labelled C.
4. Name the structure labelled D.
5. Name the structure labelled E.

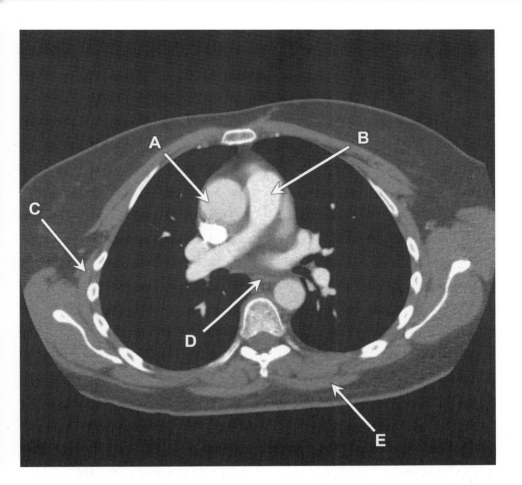

Case 8

1. Name the structure labelled A.
2. Name the structure labelled B.
3. Name the structure labelled C.
4. Name the structure labelled D.
5. Name the structure labelled E.

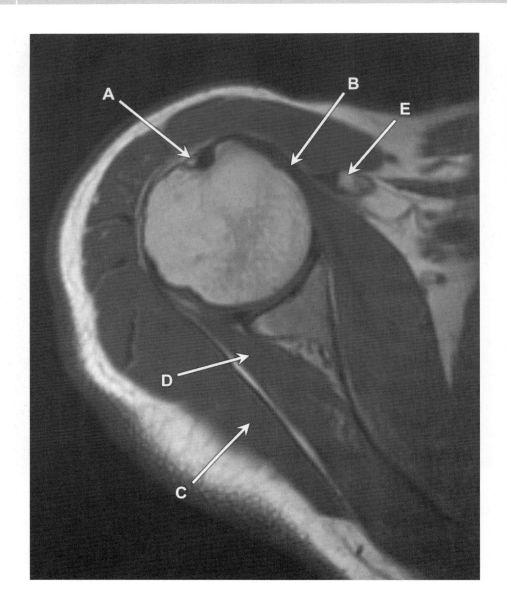

Case 9

1. Name the structure labelled A.
2. Name the structure labelled B.
3. Name the structure labelled C.
4. Name the structure labelled D.
5. Name the structure labelled E.

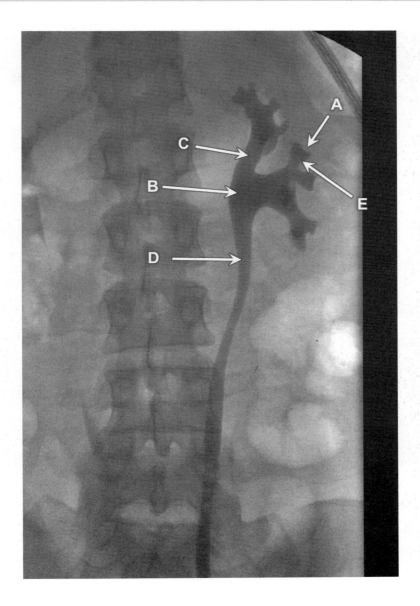

Case 10

1. Name the structure labelled A.
2. Name the structure labelled B.
3. Name the structure labelled C.
4. Name the structure labelled D.
5. Name the structure labelled E.

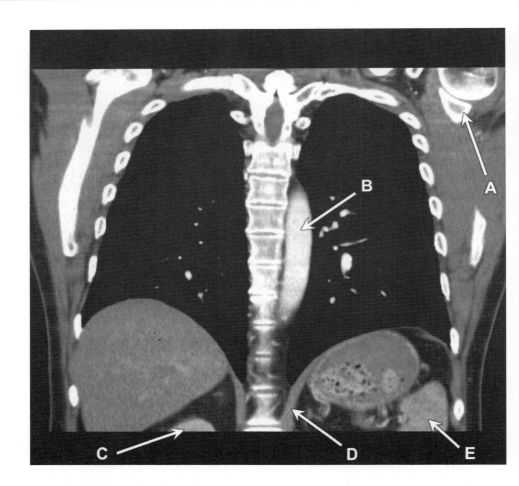

Case 11

1. Name the structure labelled A.
2. Name the structure labelled B.
3. Name the structure labelled C.
4. Name the structure labelled D.
5. Name the structure labelled E.

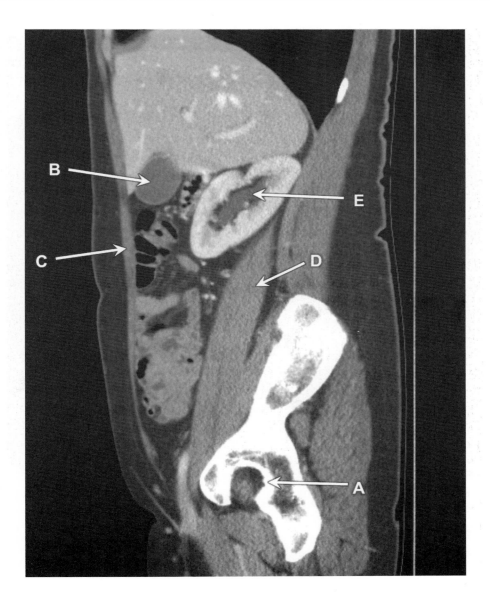

Case 12

1. Name the structure labelled A.
2. Name the structure labelled B.
3. Name the structure labelled C.
4. Name the structure labelled D.
5. Name the structure labelled E.

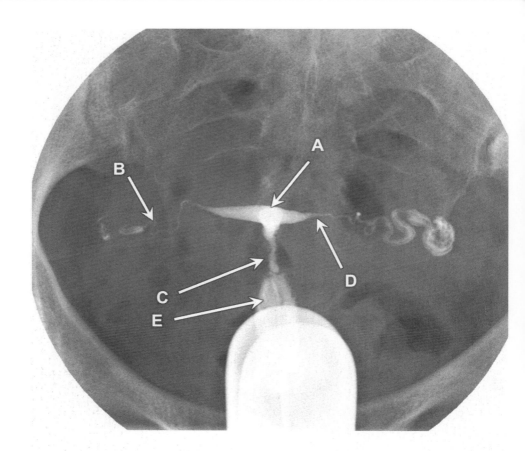

Case 13

1. Name the structure labelled A.
2. Name the structure labelled B.
3. Name the structure labelled C.
4. Name the structure labelled D.
5. Name the structure labelled E.

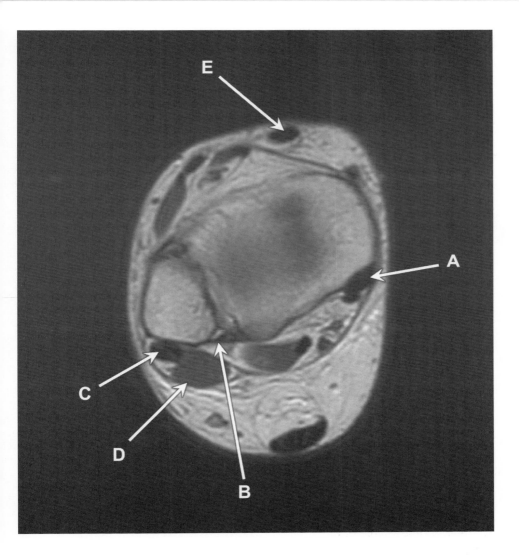

Case 14

1. Name the structure labelled A.

2. Name the structure labelled B.

3. Name the structure labelled C.

4. Name the structure labelled D.

5. Name the structure labelled E.

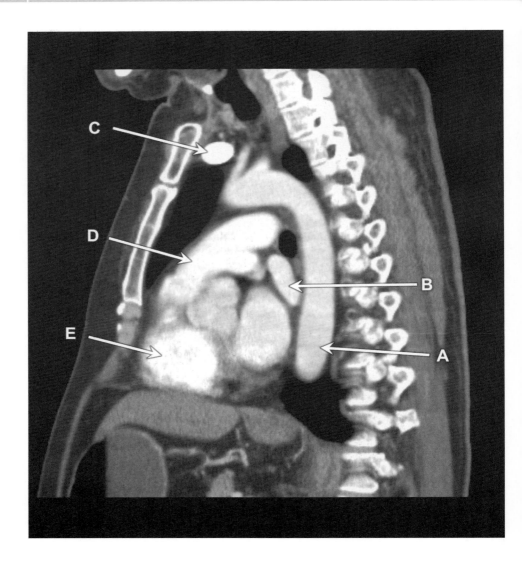

Case 15

1. Name the structure labelled A.

2. Name the structure labelled B.

3. Name the structure labelled C.

4. Name the structure labelled D.

5. Name the structure labelled E.

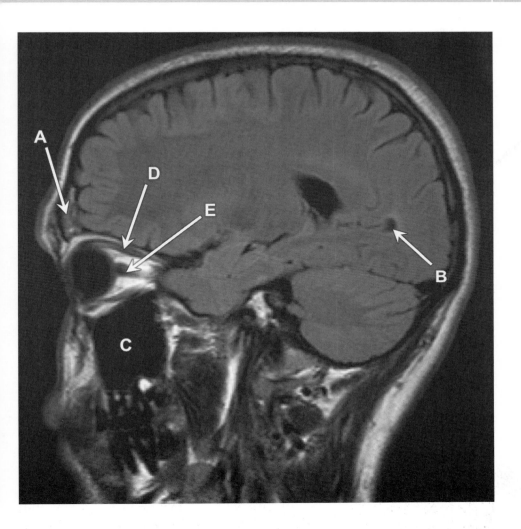

Case 16

1. Name the structure labelled A.
2. Name the structure labelled B.
3. Name the structure labelled C.
4. Name the structure labelled D.
5. Name the structure labelled E.

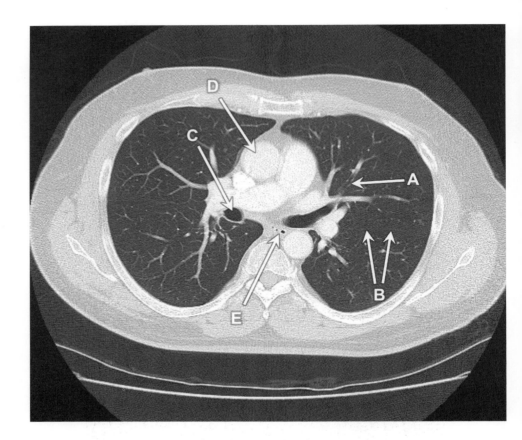

Case 17

1. Name the structure labelled A.
2. Name the structure labelled B.
3. Name the structure labelled C.
4. Name the structure labelled D.
5. Name the structure labelled E.

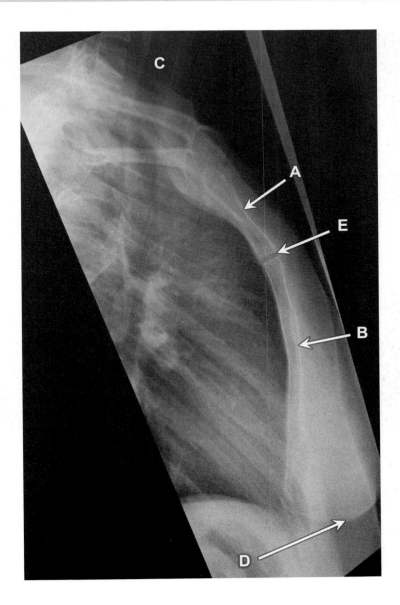

Case 18

1. Name the structure labelled A.
2. Name the structure labelled B.
3. Name the structure labelled C.
4. Name the structure labelled D.
5. Name the structure labelled E.

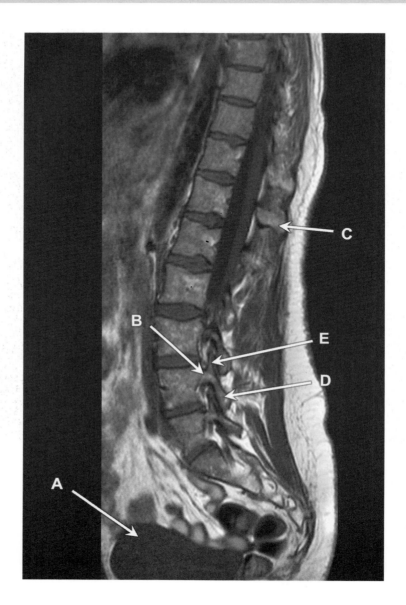

Case 19

1. Name the structure labelled A.
2. Name the structure labelled B.
3. Name the structure labelled C.
4. Name the structure labelled D.
5. Name the structure labelled E.

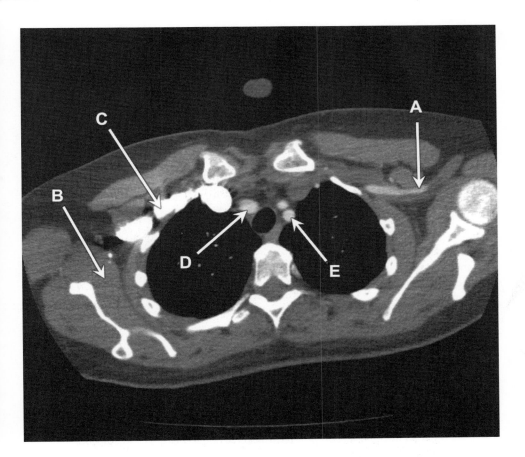

Case 20

1. Name the structure labelled A.
2. Name the structure labelled B.
3. Name the structure labelled C.
4. Name the structure labelled D.
5. Name the structure labelled E.

Case 1

AP radiograph. Pelvis.

1. Lesser trochanter of the right femur
2. Left sacral foramen of S3 vertebra
3. Fovea capitis of the left femur
4. Right pelvic brim
5. Caecum

Case 2

CT aortogram, abdomen. Axial section.

1. Abdominal aorta
2. Gallbladder
3. Coeliac trunk
4. Common hepatic artery
5. Splenic artery

There are several normal variants of the arterial anatomy of the coeliac axis and superior mesenteric artery. Candidates are advised to familiarize themselves with these.

Case 3

MRI cerebral aqueduct. Sagittal section.

1. Pericallosal artery
2. Pons
3. Basilar artery
4. Cisterna magna
5. Interpeduncular cistern

The basilar artery often takes a slightly tortuous course and is, therefore, seen in cross section on this sagittal image.

Case 4

CT chest/ abdomen. Axial section.

1. Right latissimus dorsi muscle
2. Segment 4(A)
3. Inferior vena cava
4. Oesophagus
5. Xiphoid process of sternum (xiphisternum)

Case 5

Plain radiograph, lateral cervical spine.

1. Soft palate
2. Angle of the mandible (not possible to say which side)
3. Inferior articular process of C6 vertebra
4. Posterior clinoid process
5. Superior articular process of C7 vertebra

Case 6

CT abdomen. Coronal section.

1. Right external iliac artery
2. Splenic vein
3. Right common femoral vein
4. Caecum
5. Portal vein

Remember that the blood vessels change their name from 'external iliac' to 'femoral' as they pass the inguinal ligament.

Case 7

CT facial bones. Axial section.

1. Right foramen transversarium (of CI vertebra)
2. Left zygomatic bone (zygomatic process)
3. Left styloid process
4. Dens (odontoid process of the C2 vertebra)
5. Right mandibular ramus

Case 8

CT chest. Axial section.

1. Ascending aorta
2. Pulmonary trunk
3. Right serratus anterior muscle
4. Oesophagus
5. Left trapezius muscle

Case 9

MRI shoulder. T1W axial section.

1. Transverse humeral ligament
2. Tendon of the subscapularis muscle
3. Deltoid muscle
4. Infraspinatous muscle
5. Coracoid process of the scapula

The transverse humeral ligament spans the biceps groove and is inseparable from the distal subscapularis muscle. At this level, the muscle posterior to the scapula must be infraspinatus (supraspinatous is much higher).

Case 10

Nephrogram, left kidney.

1. Renal papilla
2. Renal pelvis
3. Major calyx (upper pole)
4. Ureter
5. Minor calyx

Case 11

CT chest. Coronal section.

1. Left glenoid
2. Descending thoracic aorta
3. Right kidney (upper pole)
4. Left diaphragmatic crus
5. Spleen

This might look tricky at first but these should all be familiar structures, just seen in an unfamiliar way.

Case 12

CT abdomen. Parasagittal section.

1. Right acetabulum
2. Gallbladder
3. Right rectus abdominis muscle
4. Right psoas muscle
5. Right renal pelvis

This is the right side as liver and gallbladder are on the image!

Case 13

Hysterosalpingogram.

1. Fundus of the uterus
2. Right uterine (Fallopian) tube
3. Cervix
4. Isthmus of the left uterine (Fallopian) tube
5. Vagina

Case 14

MRI ankle. T1W axial section.

1. Tibialis posterior tendon
2. Posterior tibiofibular ligament
3. Peroneus longus tendon
4. Peroneus brevis muscle
5. Tibialis anterior tendon

Case 15

CT chest. Parasagittal section.

1. Descending thoracic aorta
2. Left pulmonary vein
3. Left brachiocephalic vein
4. Pulmonary trunk
5. Right ventricle

B is difficult. It is straightforward to work out if not immediately obvious: a structure travelling horizontally behind the heart; we are left of the midline as we can see the descending aorta; it is not connected to the pulmonary arterial tree…. it must be the left pulmonary vein!

Case 16

MRI head. T1W parasagittal section.

1. Frontal sinus
2. Occipital horn of the lateral ventricle
3. Maxillary sinus
4. Superior rectus muscle
5. Optic nerve

It is not possible to determine the side (left or right) based on the image.

Case 17

CT chest (lung windows). Axial section.

1. Lingular bronchus
2. Left oblique fissure
3. Right main bronchus
4. Ascending aorta
5. Oesophagus

Note that 'A' is hollow so is an airway, not a vessel.

Case 18

Plain radiograph. Lateral sternum.

1. Manubrium of sternum
2. Body of sternum
3. Trachea
4. Breast
5. Manubriosternal joint

Case 19

MRI lumbar spine. T1W sagittal section.

1. Uterus
2. Pedicle of L4 vertebra
3. Spinous process of T12 vertebra
4. Inferior articular process of L4 vertebra
5. Superior articular process of L4 vertebra

Case 20

CT chest. Axial section.

1. Left axillary artery
2. Right subscapularis muscle
3. Right subclavian vein
4. Brachiocephalic trunk
5. Left subclavian artery

The right subclavian vein is filled with dense contrast giving the streak artefact seen.

Of the two vessels in the left axilla, one would expect there to be contrast in the artery, not the vein, at this phase of the scan.

You have 75 minutes to complete the examination

Difficulty rating: Kilimanjaro

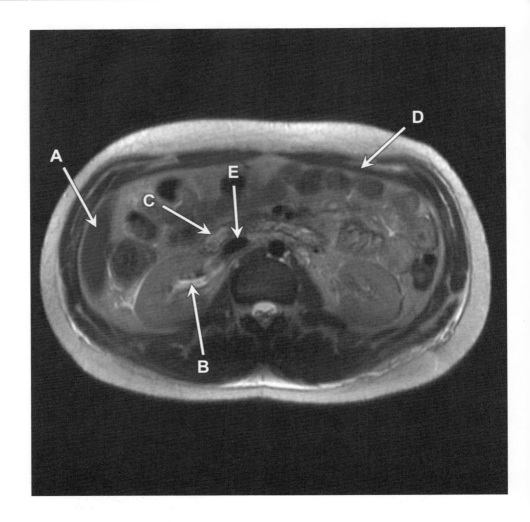

Case 1

1. Name the structure labelled A.
2. Name the structure labelled B.
3. Name the structure labelled C.
4. Name the structure labelled D.
5. Name the structure labelled E.

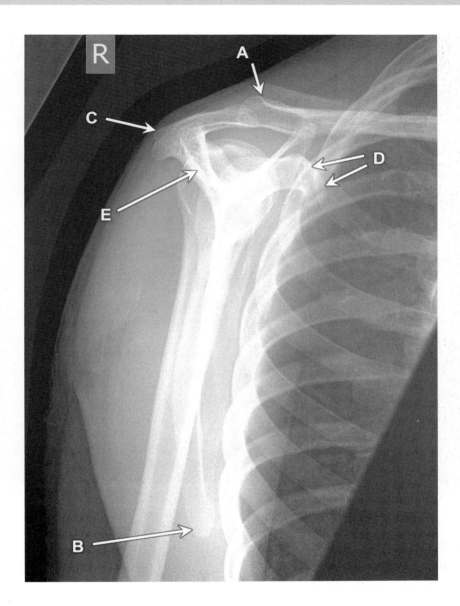

Case 2

1. Name the structure labelled A.

2. Name the structure labelled B.

3. Name the structure labelled C.

4. Name the structure labelled D.

5. Name the structure labelled E.

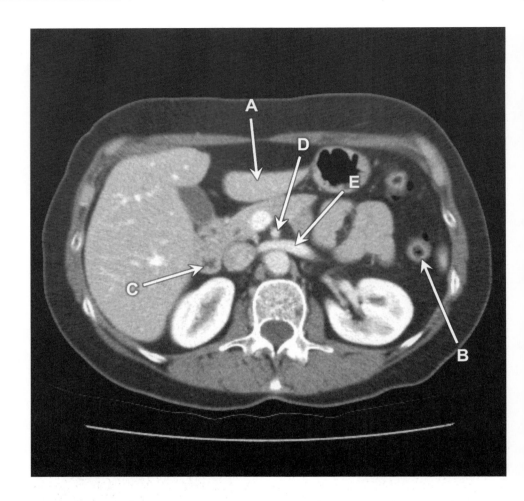

Case 3

1. Name the segment of the liver labelled A.
2. Name the structure labelled B.
3. Name the structure labelled C.
4. Name the structure labelled D.
5. Name the structure labelled E.

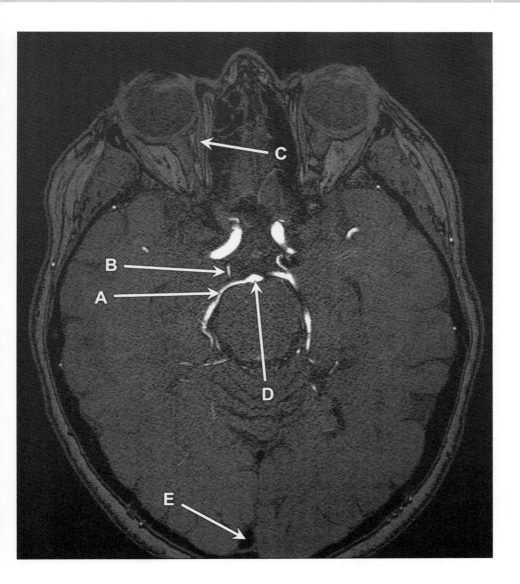

Case 4

1. Name the structure labelled A.
2. Name the structure labelled B.
3. Name the structure labelled C.
4. Name the structure labelled D.
5. Name the structure labelled E.

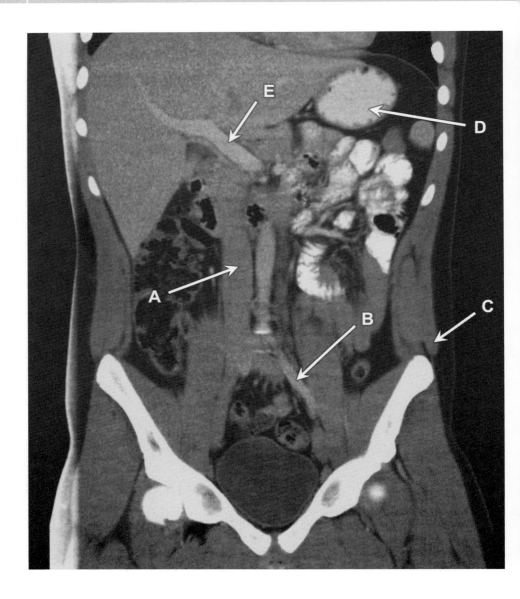

Case 5

1. Name the structure labelled A.

2. Name the structure labelled B.

3. Name the structure labelled C.

4. Name the structure labelled D.

5. Name the structure labelled E.

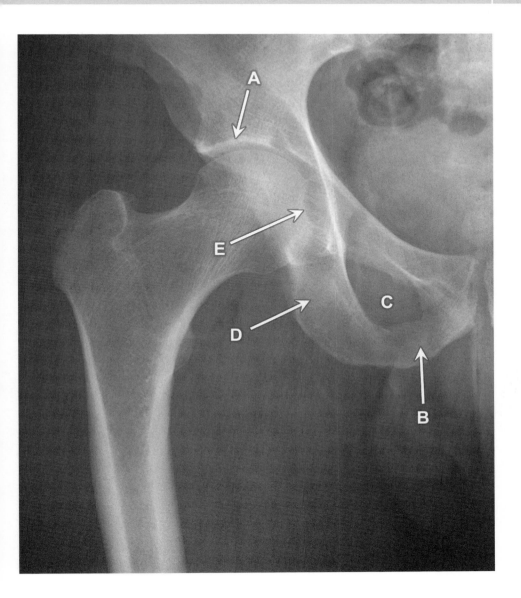

Case 6

1. Name the structure labelled A.
2. Name the structure labelled B.
3. Name the structure labelled C.
4. Name the structure labelled D.
5. Name the structure labelled E.

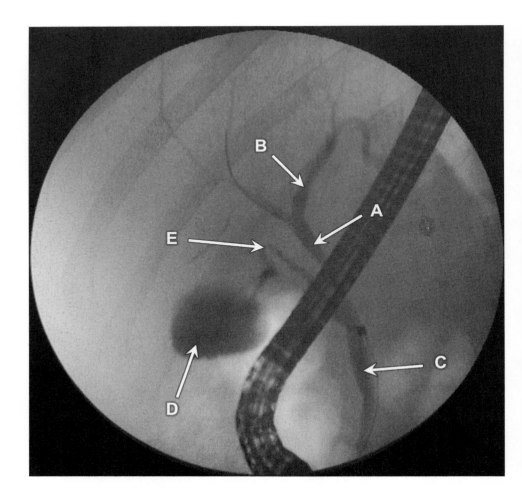

Case 7

1. Name the structure labelled A.

2. Name the structure labelled B.

3. Name the structure labelled C.

4. Name the structure labelled D.

5. Name the structure labelled E.

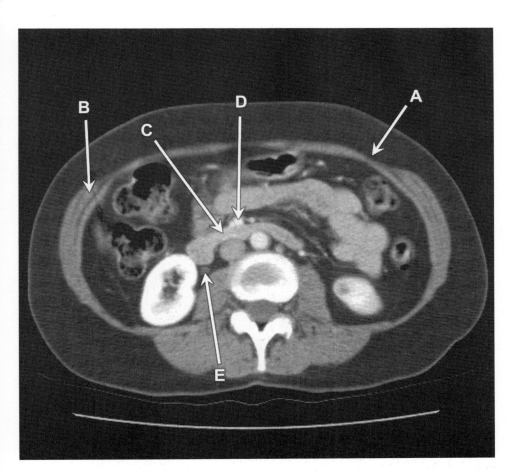

Case 8

1. Name the craniocaudal 'line' labelled A.
2. Name the structure labelled B.
3. Name the structure labelled C.
4. Name the structure labelled D.
5. Name the structure labelled E.

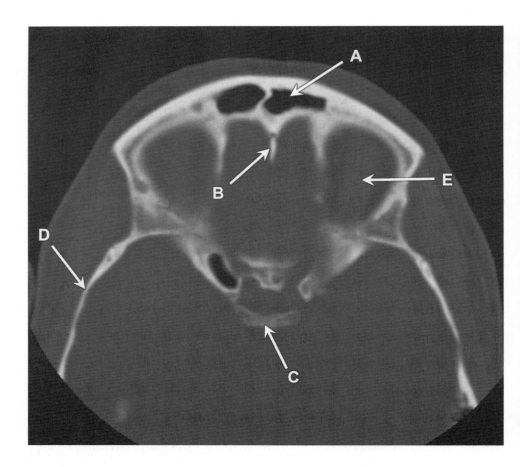

Case 9

1. Name the structure labelled A.
2. Name the structure labelled B.
3. Name the structure labelled C.
4. Name the structure labelled D.
5. Name the structure labelled E.

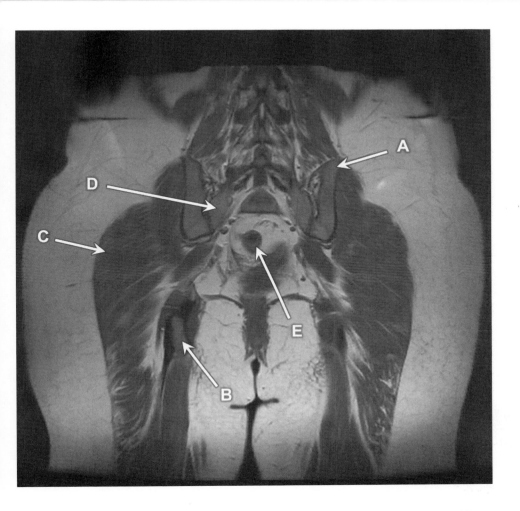

Case 10

1. Name the structure labelled A.
2. Name the structure labelled B.
3. Name the structure labelled C.
4. Name the structure labelled D.
5. Name the structure labelled E.

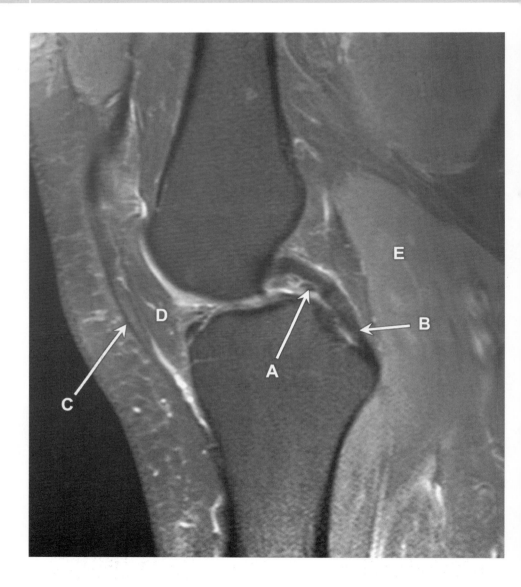

Case 11

1. Name the structure labelled A.
2. Name the structure labelled B.
3. Name the structure labelled C.
4. Name the structure that occupies the space labelled D.
5. Name the structure labelled E.

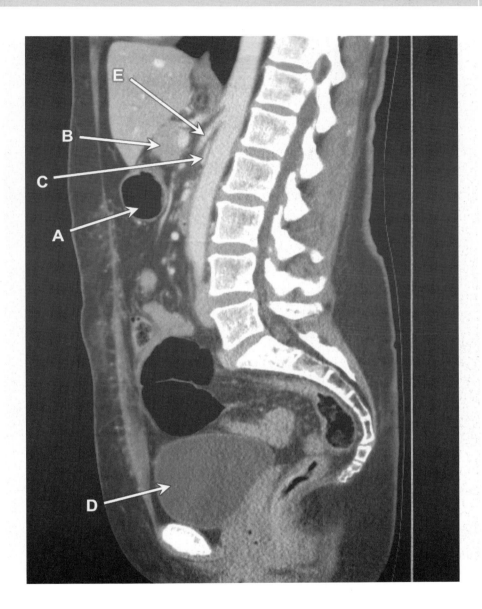

Case 12

1. Name the structure labelled A.
2. Name the structure labelled B.
3. Name the structure labelled C.
4. Name the structure labelled D.
5. Name the structure labelled E.

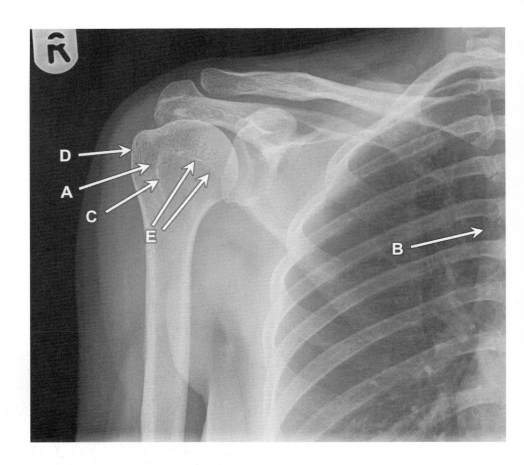

Case 13

1. Name the structure labelled A.
2. Name the structure labelled B.
3. Name the structure labelled C.
4. Name the structure labelled D.
5. Name the structure labelled E.

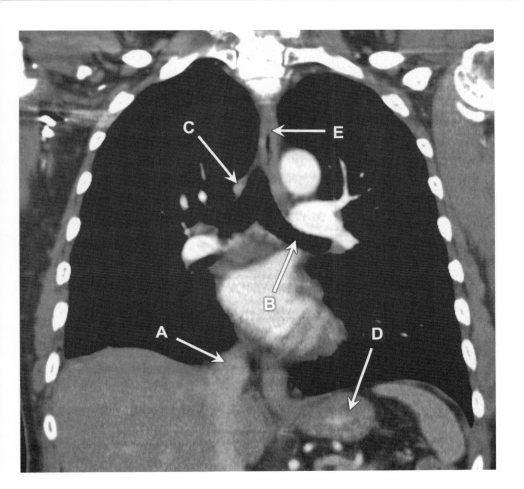

Case 14

1. Name the structure labelled A.
2. Name the structure labelled B.
3. Name the structure labelled C.
4. Name the structure labelled D.
5. Name the structure labelled E.

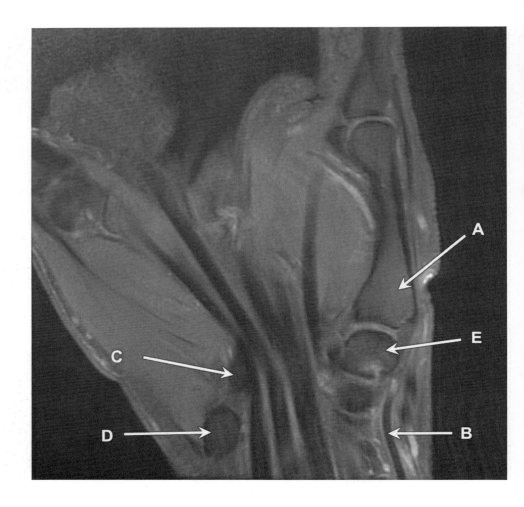

Case 15

1. Name the structure labelled A.
2. Name the structure labelled B.
3. Name the structure labelled C.
4. Name the structure labelled D.
5. Name the structure labelled E.

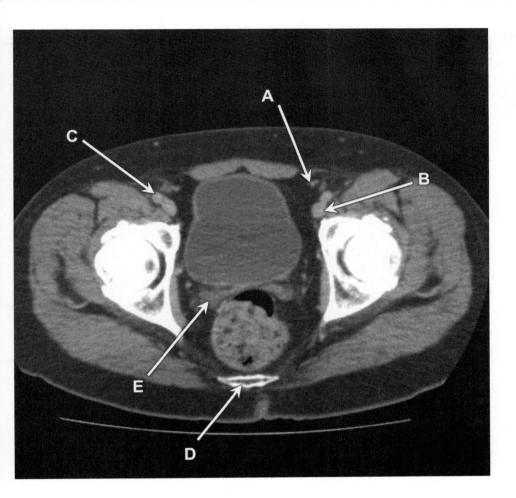

Case 16

1. Name the two rounded structures labelled A.
2. Name the structure labelled B.
3. Name the structure labelled C.
4. Name the structure labelled D.
5. Name the structure labelled E.

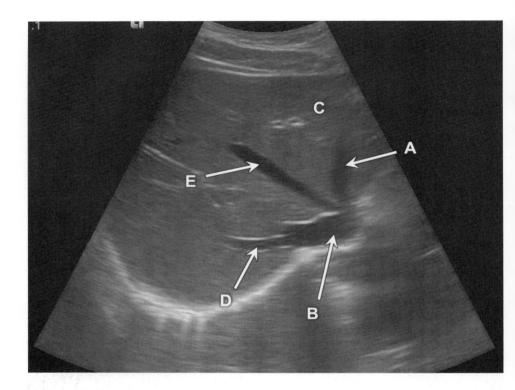

Case 17

1. Name the structure labelled A.
2. Name the structure labelled B.
3. Name the lobe of the liver labelled C.
4. Name the structure labelled D.
5. Name the structure labelled E.

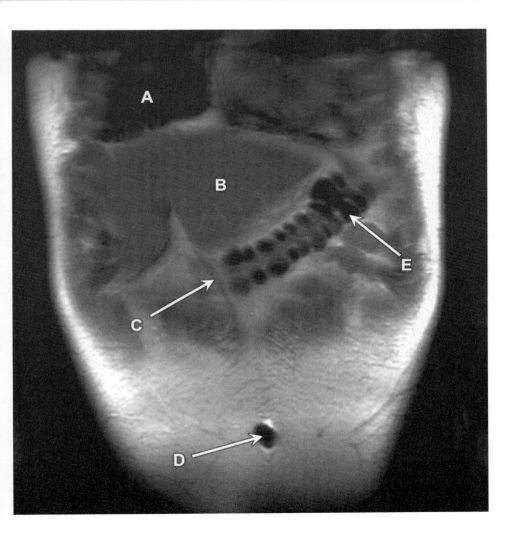

Case 18

1. Name the structure labelled A.

2. Name the structure labelled B.

3. Name the structure labelled C.

4. Name the structure labelled D.

5. Name the structure labelled E.

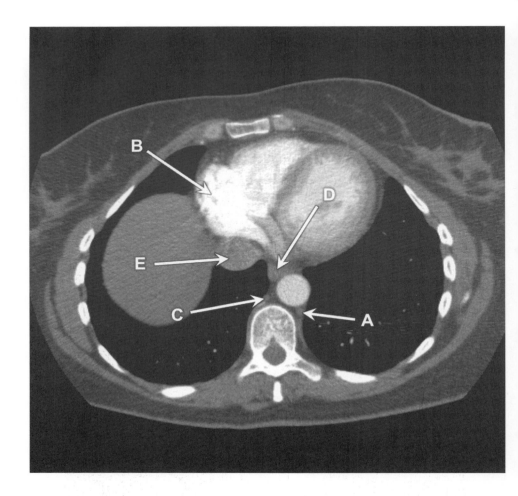

Case 19

1. Name the structure labelled A.
2. Name the structure labelled B.
3. Name the structure labelled C.
4. Name the structure labelled D.
5. Name the structure labelled E.

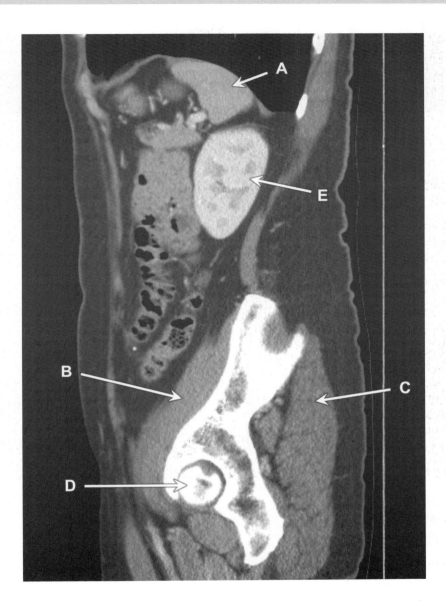

Case 20

1. Name the structure labelled A.
2. Name the structure labelled B.
3. Name the structure labelled C.
4. Name the structure labelled D.
5. Name the structure labelled E.

Case 1

MRI abdomen. T2W axial section

1. Liver
2. Right renal pelvis
3. Duodenum (second part)
4. Left transversus abdominis muscle
5. Inferior vena cava

The blood vessels show flow void artefact and are black on this image. B, therefore is not a vessel.

Case 2

Plain radiograph. Shoulder 'Y' view.

1. Right clavicle
2. Inferior angle of the right scapula
3. Right acromium
4. Right coracoid process of the scapula
5. Right scapular spine

Case 3

CT abdomen. Axial section

1. Segment 3
2. Descending colon
3. Duodenum (second part)
4. Superior mesenteric artery
5. Left renal vein

The candidate should be familiar with the liver segments. Segments 2 and 3 make up the left lobe, 2 superiorly and 3 inferiorly, divided at the level of the portal vein.

Case 4

MR-angio. Circle of Willis.

1. Right posterior cerebral artery
2. Right posterior communicating artery
3. Right medial rectus muscle
4. Basilar artery
5. Superior sagittal sinus

Case 5

CT abdomen. Coronal section.

1. Inferior vena cava
2. Left common iliac artery
3. Left external oblique muscle
4. Stomach
5. Portal vein

Case 6

Plain radiograph. Right hip.

1. Right acetabulum
2. Right inferior pubic ramus
3. Right obturator foramen
4. Right ischium
5. Right fovea capitis

One should remember that the inferior portion of the obturator ring is made up of the ischium laterally and the inferior ramus of the pubic bone medially. It is easy to mistake the whole structure for the inferior pubic ramus.

Case 7

ERCP.

1. Common hepatic duct
2. Left hepatic duct
3. Common bile duct
4. Gallbladder
5. Cystic duct

Case 8

CT abdomen. Axial section.

1. Linea semilunaris
2. Right internal oblique muscle
3. Duodenum (third part)
4. Superior mesenteric artery
5. Right ureter

The linea semilunaris is also known as 'Spiegel's line' and is the site of Spiegelian herniae (through the Spiegelian fascia).

Case 9

CT orbits. Axial section.

1. Left frontal sinus
2. Crista galli
3. Dorsum sellae
4. Squamous part of the right temporal bone
5. Left superior rectus muscle

Case 10

MRI pelvis. T1W coronal section.

1. Left iliac crest
2. Right ischial tuberosity
3. Right gluteus maximus muscle
4. Right sacral ala
5. Rectum

Case 11

MRI knee. T2W sagittal section.

1. Meniscofemoral ligament (of Humphrey)
2. Posterior cruciate ligament
3. Patellar retinaculum
4. Hoffa's fat pad
5. Medial head of gastrocnemius

The meniscofemoral ligament (when visible) has two main variants and names: when passing anterior to the posterior cruciate ligament, it is called the ligament of Humphrey; when passing posterior to the posterior cruciate, it is called the ligament of Wrisberg.

We know the gastrocnemius seen is the medial head as the posterior cruciate lies just medial to the midline of the knee.

Case 12

CT abdomen. Sagittal section.

1. Stomach
2. Pancreas
3. Left renal vein
4. Urinary bladder
5. Superior mesenteric artery

'A' might be mistaken for the transverse colon but this is seen lying much lower in the abdomen near the distended loops of sigmoid. 'C' should now be familiar as the flattened left renal vein passing over the abdominal aorta. 'B' might be mistaken for a filled duodenum but this structure is clearly solid and the splenic vein can be seen at the posterior aspect.

Case 13

Plain radiograph. AP view of right shoulder.

1. Bicipital groove
2. Superior vena cava
3. Lesser tuberosity of the humerus
4. Greater tuberosity of the humerus
5. Anatomical neck of the humerus

The 'surgical neck' of the humerus is the region at the proximal diaphysis just distal to the tuberosities. This is the most common site of fractures.

Case 14

CT chest. Coronal section.

1. Inferior vena cava
2. Left main bronchus
3. Azygos vein
4. Stomach
5. Oesophagus

Case 15

MRI wrist. Volar aspect. Coronal section.

1. First (or thumb) metacarpal bone
2. Radial artery
3. Hook of the hamate
4. Pisiform bone
5. Trapezium bone

Identifying this as the wrist may be the most difficult part; once this is done, the rest is straightforward. There is nowhere else in the body where so many tendons come together through such a narrow tunnel (and the base of thumb should be recognizable). Note that 'B' shows flow artefact distally so is a vessel, not a tendon.

Case 16

CT pelvis. Axial section.

1. Left inferior epigastric vessels
2. Left external iliac vein
3. Right external iliac artery
4. Sacrum
5. Right seminal vesicles

Case 17

Ultrasound liver.

1. Left hepatic vein
2. Inferior vena cava
3. Left lobe of the liver
4. Right hepatic vein
5. Middle hepatic vein

Case 18

MRI abdomen. T1W coronal section.

1. Right lung
2. Left lobe of the liver
3. Falciform ligament
4. Umbilicus
5. Transverse colon

As elsewhere, once one has recognized that this is an anterior slice of the abdomen, the rest should be straightforward.

Case 19

CT chest. Axial section.

1. Hemiazygos vein
2. Right atrium
3. Azygos vein
4. Oesophagus
5. Inferior vena cava

Case 20

CT abdomen. Parasagittal section.

1. Spleen
2. Left iliacus muscle
3. Left gluteus maximus muscle
4. Left femoral head
5. Left kidney

The absence of the liver and presence of the spleen make this a left parasagittal section.

You have 75 minutes to complete the examination

Difficulty rating: North face of the Eiger

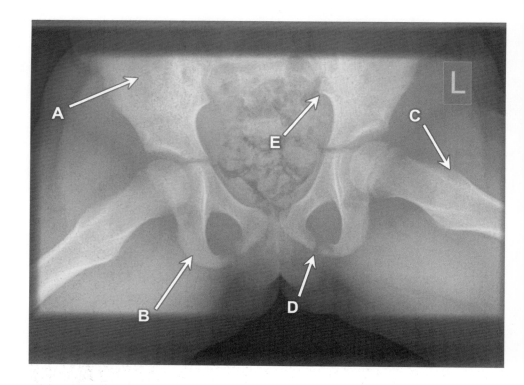

Case 1

1. Name the structure labelled A.
2. Name the structure labelled B.
3. Name the structure labelled C.
4. Name the structure labelled D.
5. Name the structure labelled E.

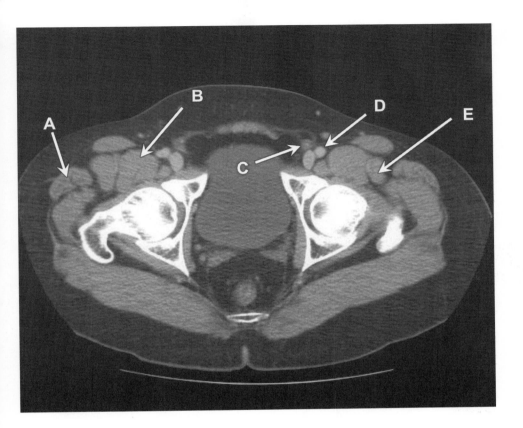

Case 2

1. Name the structure labelled A.
2. Name the structure labelled B.
3. Name the structure labelled C.
4. Name the structure labelled D.
5. Name the structure labelled E.

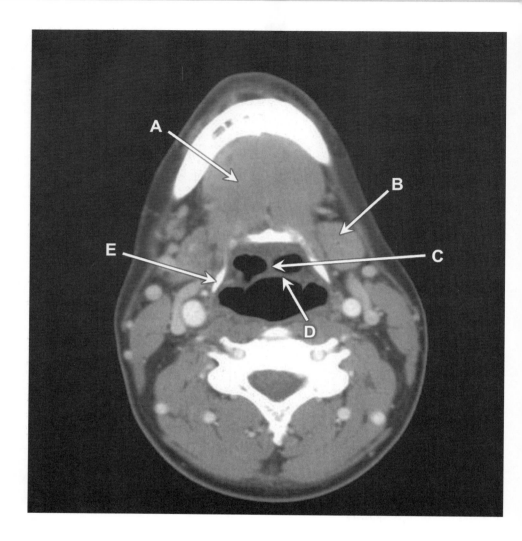

Case 3

1. Name the structure labelled A.
2. Name the structure labelled B.
3. Name the structure labelled C.
4. Name the structure labelled D.
5. Name the structure labelled E.

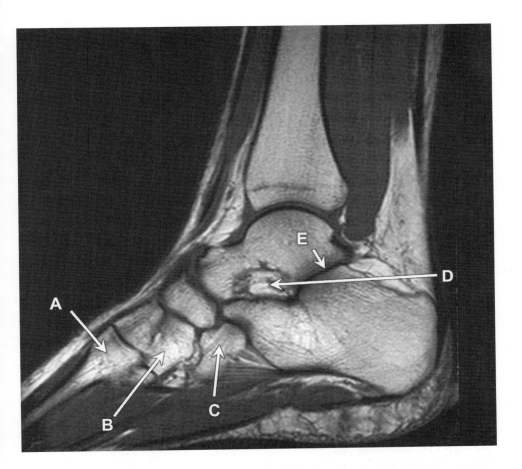

Case 4

1. Name the structure labelled A.
2. Name the structure labelled B.
3. Name the structure labelled C.
4. Name the structure labelled D.
5. Name the joint labelled E.

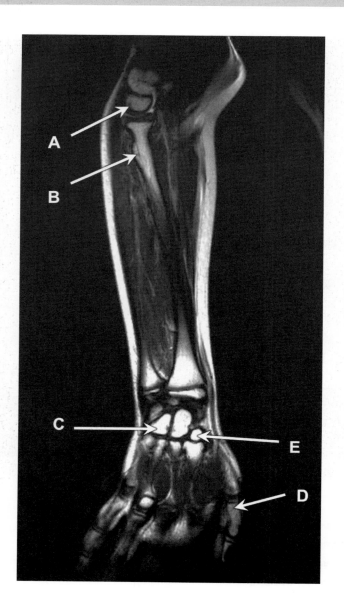

Case 5

1. Name the structure labelled A.
2. Name the structure labelled B.
3. Name the structure labelled C.
4. Name the structure labelled D.
5. Name the structure labelled E.

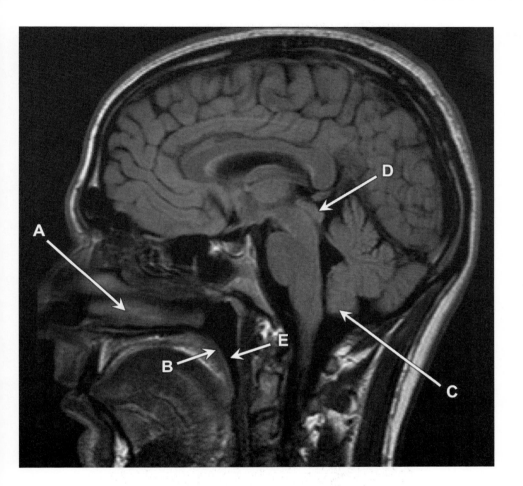

Case 6

1. Name the structure labelled A.
2. Name the structure labelled B.
3. Name the structure labelled C.
4. Name the structure labelled D.
5. Name the region labelled E.

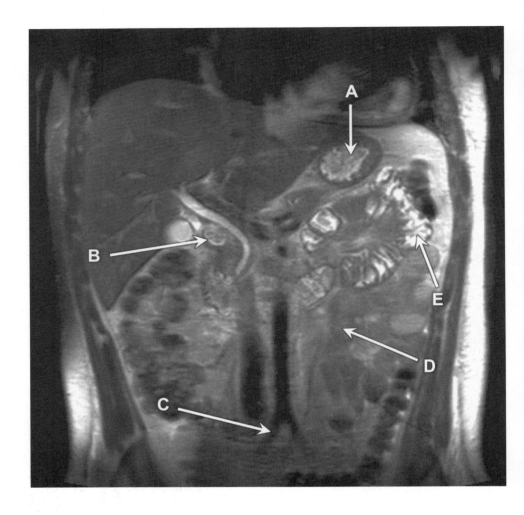

Case 7

1. Name the structure labelled A.
2. Name the structure labelled B.
3. Name the structure labelled C.
4. Name the part of the small bowel labelled D.
5. Name the part of the small bowel labelled E.

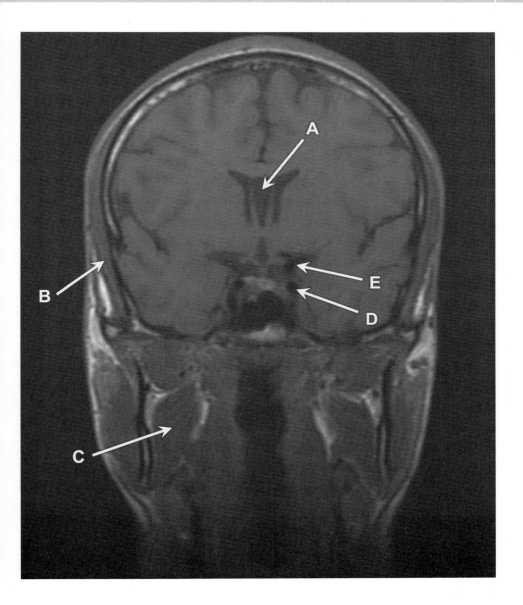

Case 8

1. Name the normal variant labelled A.

2. Name the structure labelled B.

3. Name the structure labelled C.

4. Name the structure labelled D.

5. Name the structure labelled E.

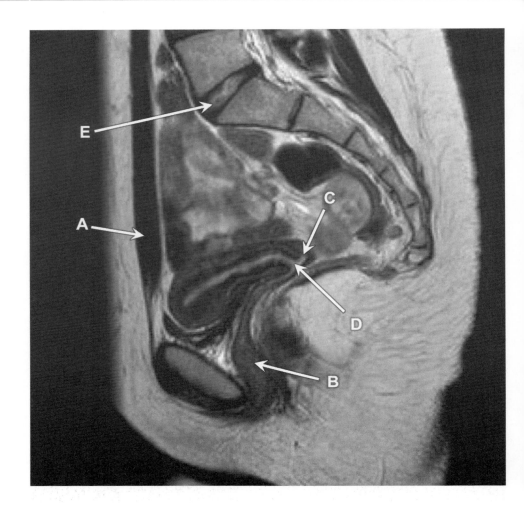

Case 9

1. Name the structure labelled A.
2. Name the structure labelled B.
3. Name the structure labelled C.
4. Name the structure labelled D.
5. Name the structure labelled E.

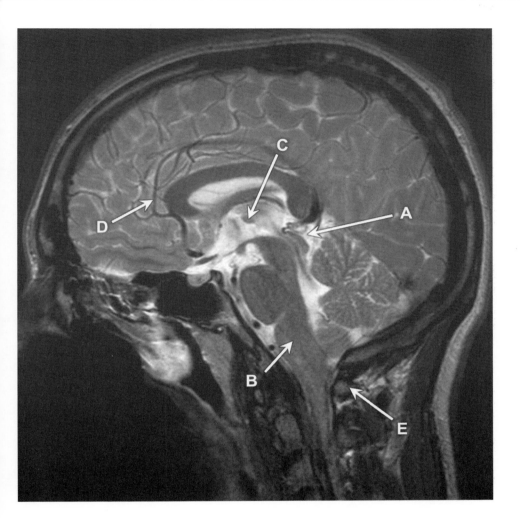

Case 10

1. Name the CSF space labelled A.
2. Name the structure labelled B.
3. Name the structure labelled C.
4. Name the vessel labelled D.
5. Name the structure labelled E.

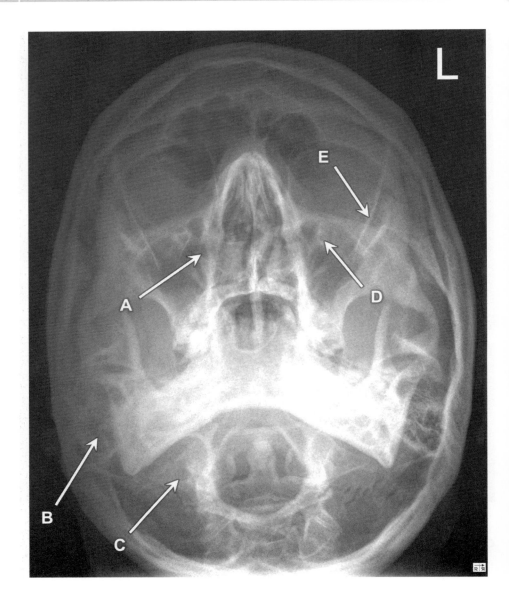

Case 11

1. Name the foramen labelled A.

2. Name the structure labelled B.

3. Name the foramen labelled C.

4. Name the foramen labelled D.

5. Name the structure labelled E.

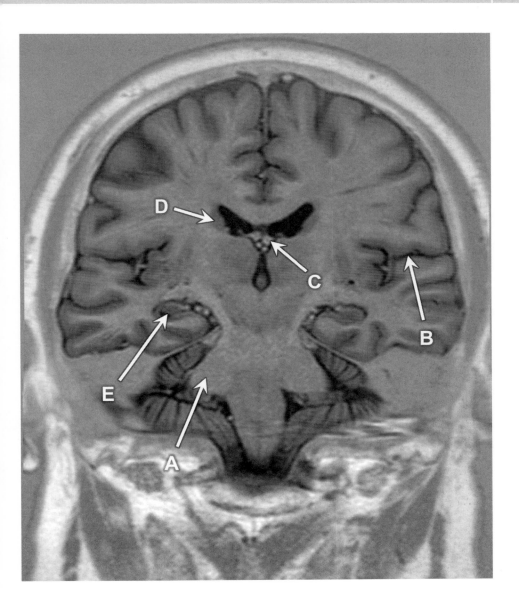

Case 12

1. Name the structure labelled A.
2. Name the structure labelled B.
3. Name the structure labelled C.
4. Name the structure labelled D.
5. Name the structure labelled E.

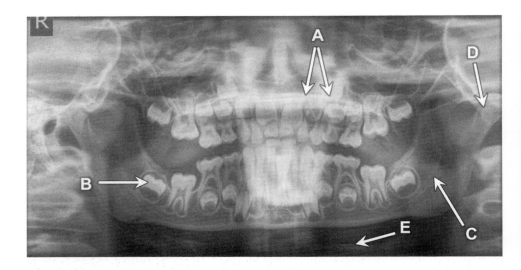

Case 13

1. Name the structure labelled A.
2. Name the structure labelled B.
3. Name the structure labelled C.
4. Name the structure labelled D.
5. Name the structure labelled E.

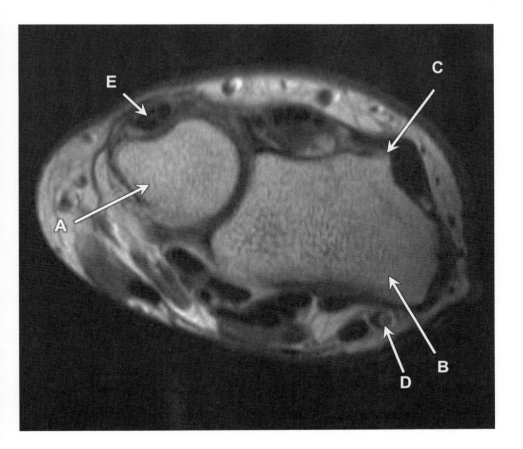

Case 14

1. Name the structure labelled A.
2. Name the structure labelled B.
3. Name the structure labelled C.
4. Name the structure labelled D.
5. Name the structure labelled E.

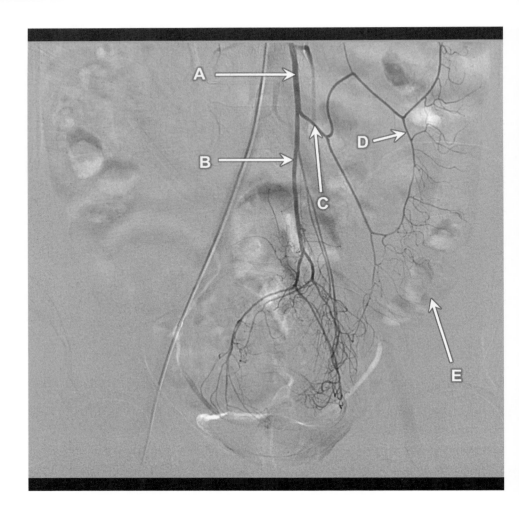

Case 15

1. Name the structure labelled A.
2. Name the structure labelled B.
3. Name the structure labelled C.
4. Name the structure labelled D.
5. Name the structure labelled E.

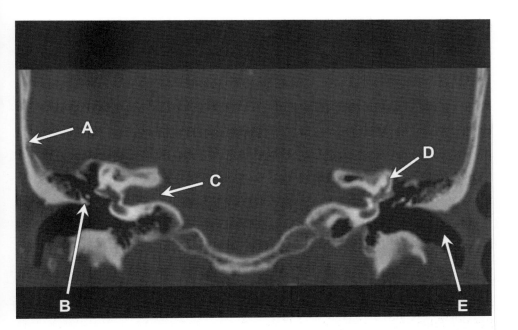

Case 16

1. Name the structure labelled A.
2. Name the structure labelled B.
3. Name the structure labelled C.
4. Name the structure labelled D.
5. Name the structure labelled E.

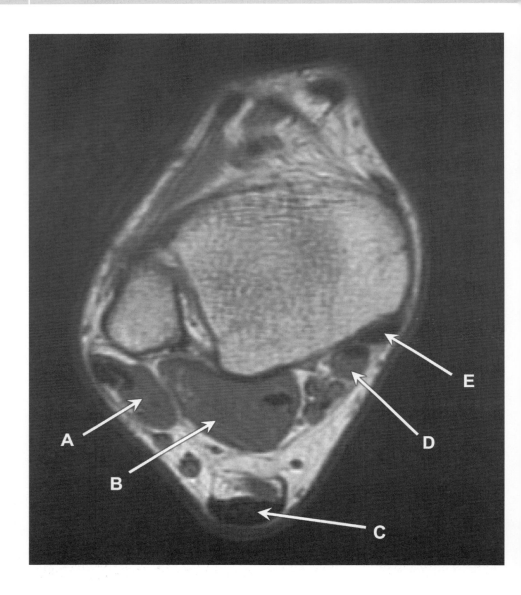

Case 17

1. Name the structure labelled A.

2. Name the structure labelled B.

3. Name the structure labelled C.

4. Name the structure labelled D.

5. Name the structure labelled E.

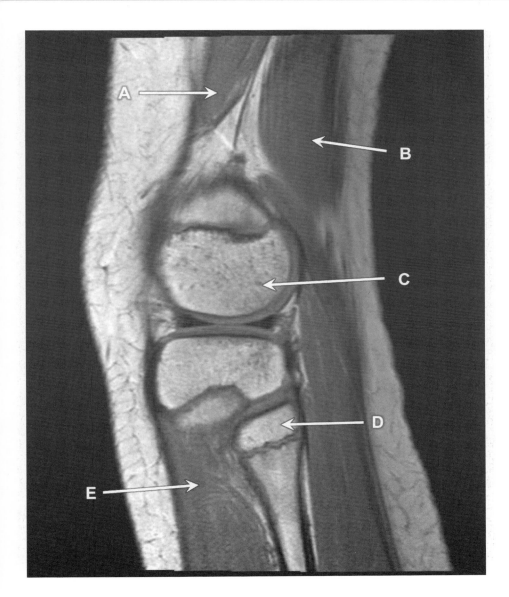

Case 18

1. Name the structure labelled A.
2. Name the structure labelled B.
3. Name the structure labelled C.
4. Name the structure labelled D.
5. Name the structure labelled E.

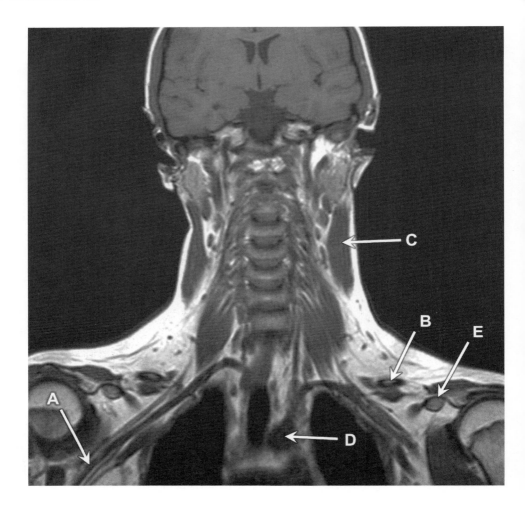

Case 19

1. Name the structure labelled A.
2. Name the structure labelled B.
3. Name the structure labelled C.
4. Name the structure labelled D.
5. Name the structure labelled E.

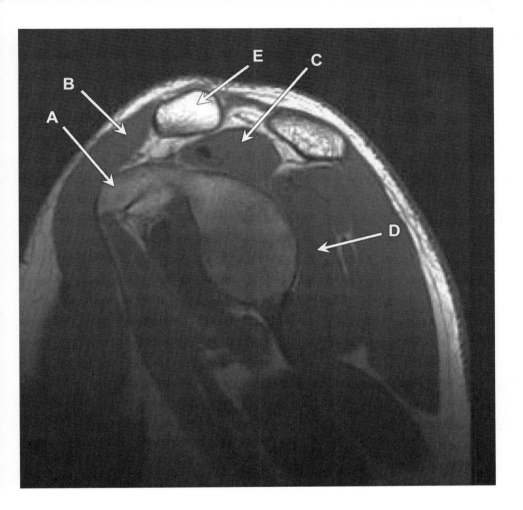

Case 20

1. Name the structure labelled A.
2. Name the structure labelled B.
3. Name the structure labelled C.
4. Name the structure labelled D.
5. Name the structure labelled E.

Case 1

Radiograph pelvis (child).

1. Right ilium
2. Right ischial ramus
3. Greater trochanter of left femur
4. Left ischiopubic syndesmosis
5. Left posterior inferior iliac spine

Case 2

CT pelvis (male) with intravenous contrast. Axial image.

1. Right tensor fascia lata muscle
2. Right iliopsoas muscle
3. Left spermatic cord
4. Left common femoral artery
5. Left rectus femoris muscle

Case 3

CT neck. Axial image.

1. Tongue
2. Left submandibular gland
3. Median glossoepiglottic fold
4. Epiglottis
5. Hyoid bone

Case 4

MRI hindfoot. T2W sagittal section.

1. Base of second metatarsal
2. Lateral cuneiform
3. Cuboid
4. Sinus tarsi/ tarsal sinus
5. Posterior talocalcaneal/ subtalar joint

Case 5

MRI forearm. T2W coronal section.

1. Capitulum of humerus
2. Tuberosity of radius
3. Hamate
4. Proximal phalanx of thumb
5. Trapezoid

Case 6

MRI brain. T1W midsagittal section.

1. Inferior turbinate
2. Soft palate
3. Cerebellar tonsil
4. Tectum of midbrain
5. Nasopharynx

Case 7

MRI abdomen. T2W coronal section.

1. Stomach
2. Duodenum
3. Right common iliac artery
4. Ileum
5. Jejunum

Case 8

MRI brain. T1W coronal section.

1. Cavum septum pellucidum
2. Right temporalis muscle
3. Right medial pterygoid muscle
4. Left internal carotid artery
5. Chiasmatic cistern

Candidates will be expected to be familiar with common normal variants such as the cavum seputum pellucidum above.

Case 9

MRI pelvis (female). T2W sagittal section.

1. Rectus abdominus muscle
2. Vagina
3. Posterior fornix of vagina
4. External os of cervix
5. L5/S1 intervertebral disc

Case 10

MRI brain. T2W sagittal section.

1. Quadrigeminal cistern
2. Medulla oblongata
3. Massa intermedia
4. Pericallosal artery
5. Posterior arch of atlas (C1)

Case 11

Occipitomental skull radiograph.

1. Right foramen rotundum
2. Right mastoid air cells
3. Right transverse foramen of C1
4. Left infraorbital foramen
5. Left innominate line/ greater wing of sphenoid

Case 12

MRI brain. Coronal section.

1. Right middle cerebellar peduncle
2. Left Sylvian fissure
3. Left fornix
4. Head of right caudate nucleus
5. Right hippocampus

Case 13

Orthopantomogram (child).

1. Hard palate
2. Right lower third molar
3. Left inferior alveolar canal
4. Left mandibular condyle
5. Hyoid bone

Case 14

MRI wrist. Axial image at the level of the distal radius.

1. Ulna
2. Radius
3. Dorsal tubercle of radius/ Lister's tubercle
4. Radial artery
5. Tendon of extensor carpi ulnaris muscle

A difficult case. The radial artery is clearly different to the other rounded structures being circular rather than ovoid on this section and showing flow void artefact rather than a solid low signal.

Case 15

Inferior mesenteric arteriogram with digital subtraction.

1. Inferior mesenteric artery
2. Superior rectal artery
3. Left colic artery
4. Marginal artery of Drummond
6. Descending colon

Digital subtraction will remove all structures that are unchanged between the pre- and postcontrast images, effectively deleting the background. Anything that moves will remain and the patient is asked to hold their breath to limit movement of the internal organs and bony skeleton. Bowel is often seen, however, as it peristalses (this can be limited by the administration of buscopan).

Case 16

CT temporal bones. Coronal reconstruction.

1. Right squamous temporal bone
2. Right scutum
3. Right internal auditory meatus
4. Left superior semicircular canal
5. Left external auditory meatus

Case 17

MRI ankle. Axial section just superior to the mortise joint.

1. Peroneus brevis muscle
2. Flexor hallucis longus muscle
3. Achilles' tendon
4. Flexor digitorum longus muscle
5. Tendon of tibialis posterior muscle

The tendons and muscles of the lower limb are not as intimidating as they first seem and should be familiar to the candidate. Consult your favourite anatomy textbook and remember the mnemonic 'Tom, Dick and Harry' for structures E, D and B respectively (Tibialis posterior, flexor Digitorum longus, flexor Hallucis longus).

Case 18

MRI knee. T1W sagittal section.

1. Vastus lateralis muscle
2. Biceps femoris muscle
3. Lateral condyle of femur
4. Head of fibula
5. Tibialis anterior muscle

Case 19

MRI neck. T1W coronal section.

1. Right axillary artery
2. Left clavicle
3. Left sternocleidomastoid muscle
4. Aortic arch
5. Left coracoid process of scapula

Case 20

MRI shoulder. Sagittal oblique T1W section.

1. Coracoid process
2. Deltoid muscle
3. Supraspinatus muscle
4. Infraspinatus muscle
5. Clavicle

You have 75 minutes to complete the examination

Difficulty rating: North face of the Eiger

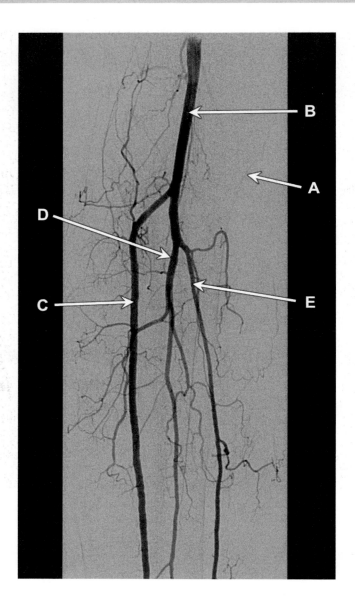

Case 1

1. Name the very faint structure labelled A.
2. Name the vessel labelled B.
3. Name the vessel labelled C.
4. Name the vessel labelled D.
5. Name the vessel labelled E.

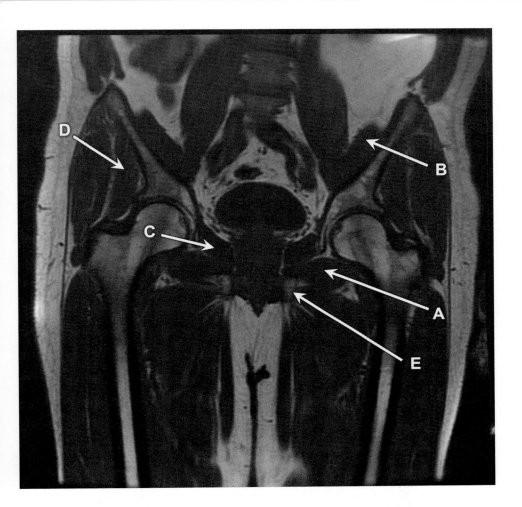

Case 2

1. Name the structure labelled A.
2. Name the structure labelled B.
3. Name the structure labelled C.
4. Name the structure labelled D.
5. Name the structure labelled E.

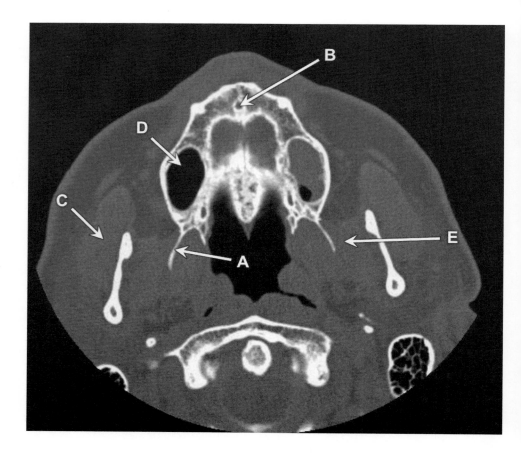

Case 3

1. Name the structure labelled A.
2. Name the foramen labelled B.
3. Name the structure labelled C.
4. Name the structure labelled D.
5. Name the structure labelled E.

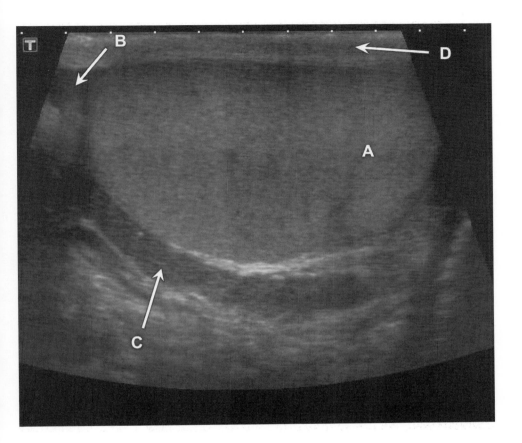

Case 4

1. Name the structure labelled A.
2. Name the structure labelled B.
3. Name the structure labelled C.
4. Name the structure labelled D.
5. Name the main group of lymph nodes that the structure labelled A drains to.

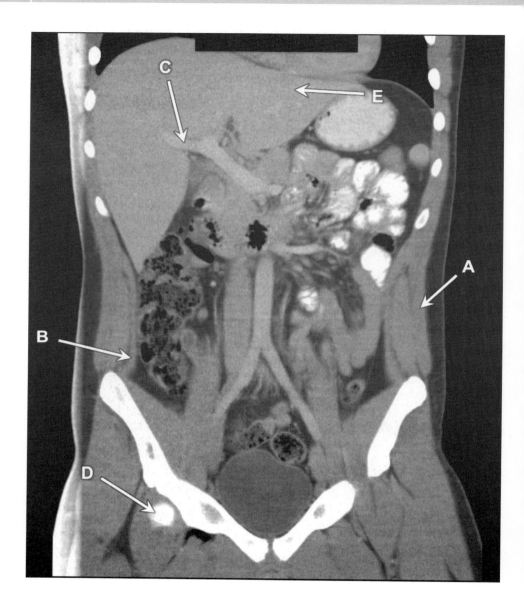

Case 5

1. Name the muscle labelled A.
2. Name the muscle labelled B.
3. Name the structure labelled C.
4. Name the structure labelled D.
5. Name the structure labelled E.

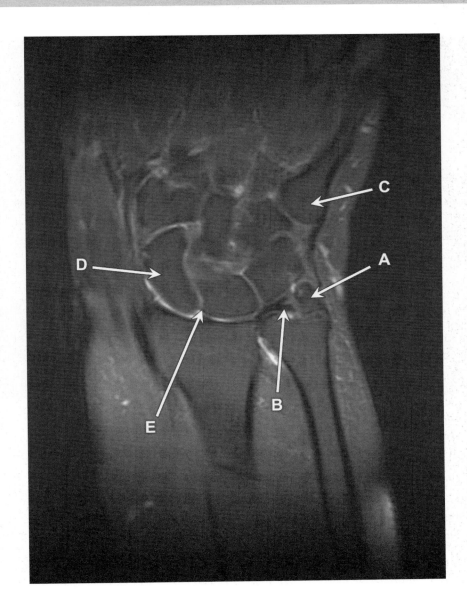

Case 6

1. Name the structure labelled A.
2. Name the structure labelled B.
3. Name the structure labelled C.
4. Name the structure labelled D.
5. Name the structure labelled E.

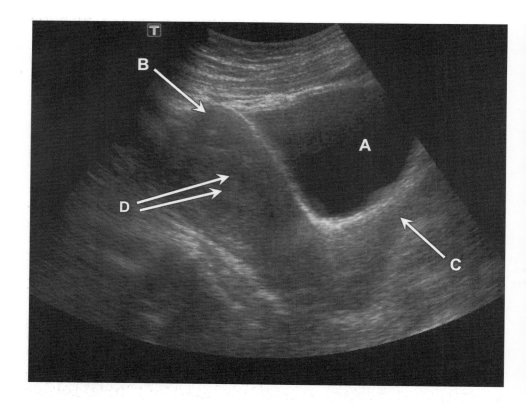

Case 7

1. Name the structure labelled A.
2. Name the structure labelled B.
3. Name the structure labelled C.
4. Name the structure labelled D.
5. Name the intraperitoneal space between structure C and the rectum.

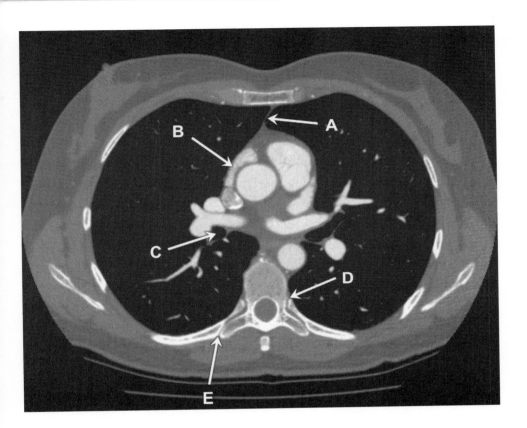

Case 8

1. Name the structure labelled A.
2. Name the structure labelled B.
3. Name the structure labelled C.
4. Name the joint labelled D.
5. Name the joint labelled E.

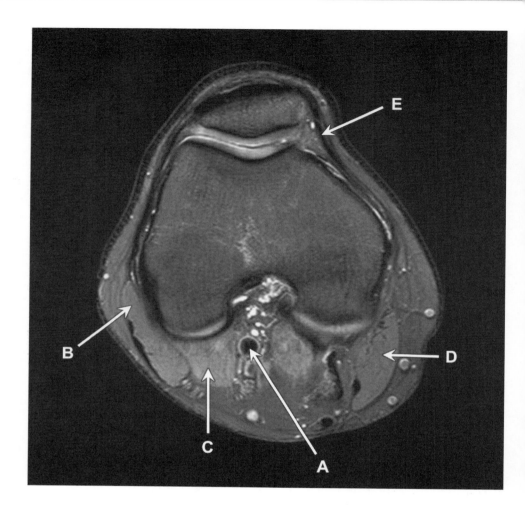

Case 9

1. Name the structure labelled A.
2. Name the structure labelled B.
3. Name the structure labelled C.
4. Name the structure labelled D.
5. Name the structure labelled E.

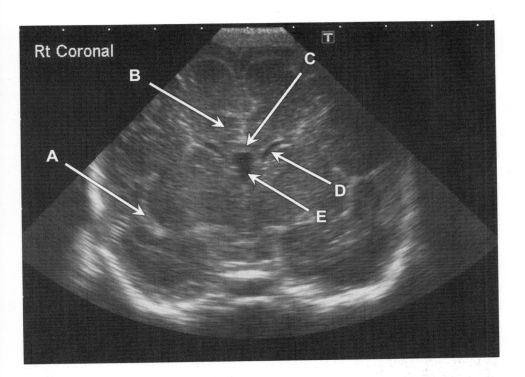

Rt Coronal

B

C

A

D

E

Case 10

1. Name the structure labelled A.

2. Name the gyrus labelled B.

3. Name the structure labelled C.

4. Name the structure labelled D.

5. Name the structure labelled E.

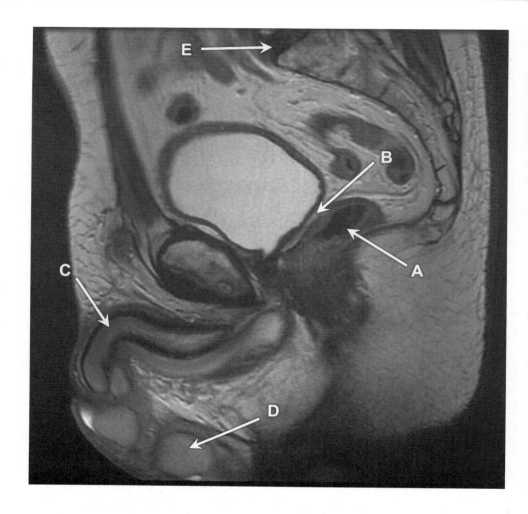

Case 11

1. Name the structure labelled A.
2. Name the space labelled B.
3. Name the structure labelled C.
4. Name the structure labelled D.
5. Name the structure labelled E.

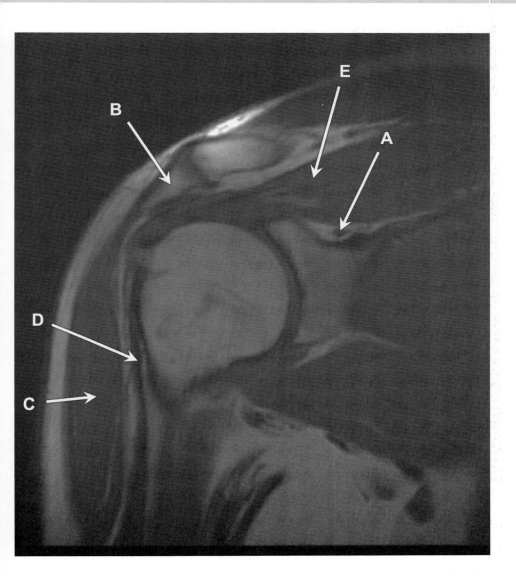

Case 12

1. Name the structure labelled A.

2. Name the structure labelled B.

3. Name the structure labelled C.

4. Name the structure labelled D.

5. Name the structure labelled E.

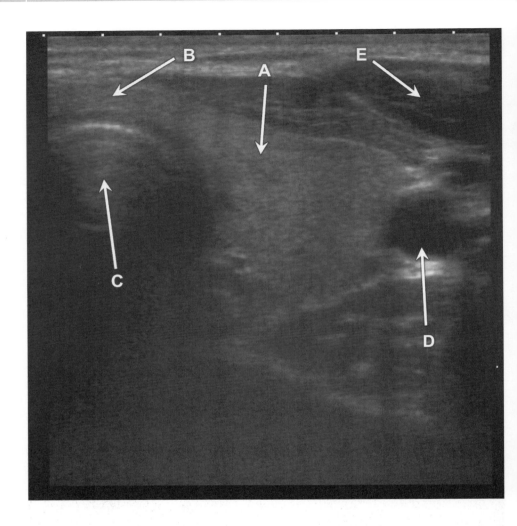

Case 13

1. Name the structure labelled A.

2. Name the structure labelled B.

3. Name the structure labelled C.

4. Name the structure labelled D.

5. Name the structure labelled E.

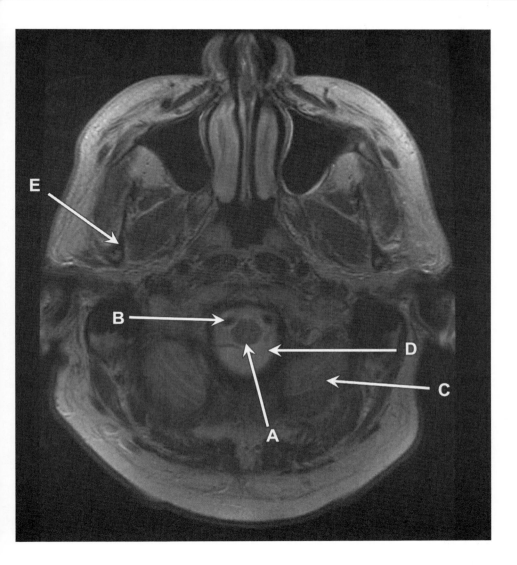

Case 14

1. Name the structure labelled A.
2. Name the structure labelled B.
3. Name the structure labelled C.
4. Name the CSF space labelled D.
5. Name the structure labelled E.

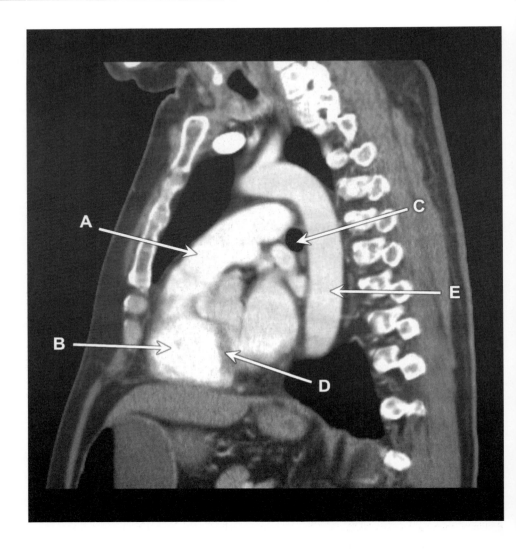

Case 15

1. Name the structure labelled A.
2. Name the structure labelled B.
3. Name the structure labelled C.
4. Name the structure labelled D.
5. At what thoracic level does the structure labelled E pass through the diaphragm?

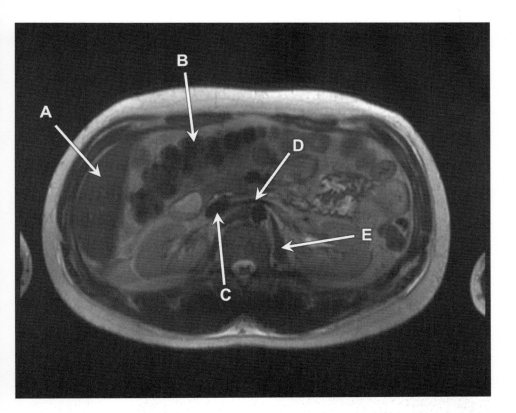

Case 16

1. Name the structure labelled A.
2. Name the structure labelled B.
3. Name the structure labelled C.
4. Name the structure labelled D.
5. Name the structure labelled E.

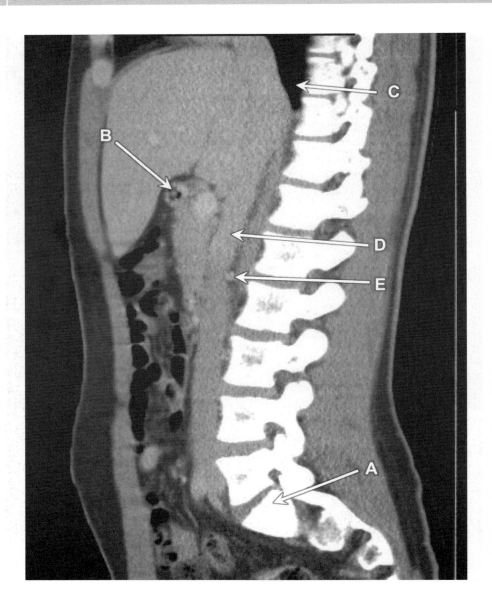

Case 17

1. Name the structure labelled A.

2. Name the structure labelled B.

3. Name the structure labelled C.

4. Name the structure labelled D.

5. Name the structure labelled E.

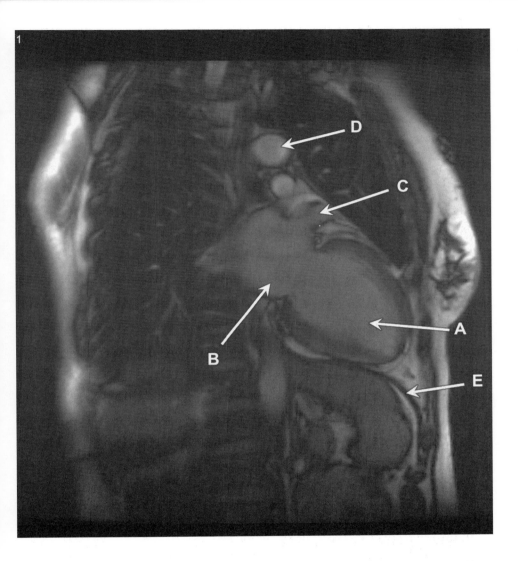

Case 18

1. Name the structure labelled A.

2. Name the structure labelled B.

3. Name the structure labelled C.

4. Name the structure labelled D.

5. Name the structure labelled E.

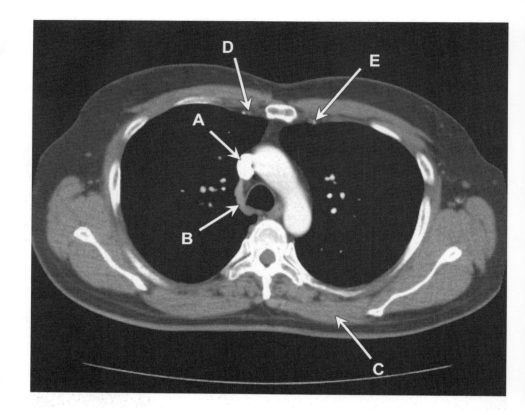

Case 19

1. Name the structure labelled A.
2. Name the structure labelled B.
3. Name the muscle labelled C.
4. Name the vessel labelled D.
5. Name the vessel labelled E.

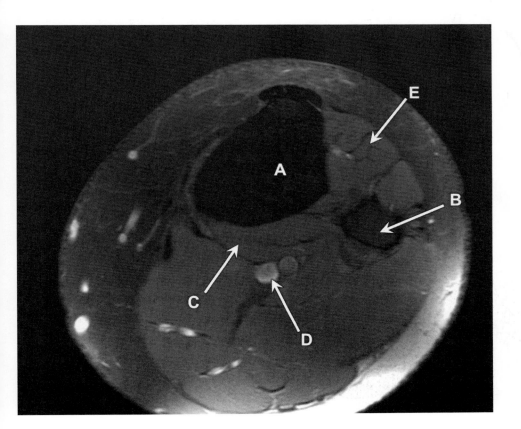

Case 20

1. Name the structure labelled A.
2. Name the structure labelled B.
3. Name the structure labelled C.
4. Name the structure labelled D.
5. Name the structure labelled E.

Case 1

Angiogram. Lower limb arteries.

1. Tibia
2. Popliteal artery
3. Anterior tibial artery
4. Posterior tibial artery
5. Peroneal artery

Case 2

MRI pelvis. Coronal T1W image.

1. Left obturator externus muscle
2. Left iliacus muscle
3. Right obturator internus muscle
4. Right gluteus minimus muscle
5. Left inferior pubic ramus

Case 3

CT base of skull. Axial imaging, 'bone windows'.

1. Right lateral pterygoid plate
2. Incisive foramen
3. Right masseter muscle
4. Right maxillary sinus
5. Left lateral pterygoid muscle

Note that, in this case, the right maxillary sinus is partly opacified by mucous. This is not an uncommon finding.

Case 4

Ultrasound testis. Longitudinal section.

1. Testis
2. Head of epididymis
3. Body of epididymis
4. Scrotal skin
5. Para-aortic lymph nodes

Case 5

CT abdomen and pelvis with intravenous contrast. Coronal reconstruction.

1. Left external oblique muscle
2. Right transversus abdominis muscle
3. Right portal vein
4. Head of right femur
5. Left lobe of liver

Case 6

MRI wrist. Coronal T2W image.

1. Ulnar styloid
2. Triangular fibrocartilaginous disc
3. Base of fifth metacarpal
4. Scaphoid
5. Scapholunate ligament

Case 7

Ultrasound pelvis (female patient). Longitudinal section.

1. Urinary bladder
2. Uterine fundus
3. Vagina
4. Endometrium
5. Pouch of Douglas

The radiologist should be familiar with the name and position of the various intraperitoneal spaces.

Case 8

CT thorax with intravenous contrast. Axial section.

1. Anterior junction line
2. Right atrium
3. Bronchus intermedius
4. Left costovertebral joint
5. Right costotransverse joint

Case 9

MRI knee. T2W axial image.

1. Popliteal artery
2. Biceps femoris muscle
3. Lateral head of gastrocnemius muscle
4. Sartorius muscle
5. Medial patellofemoral ligament/ medial retinaculum

You should beware ghosting artefacts on MR images. These are displaced reduplications of some or part of the image and occur in the phase encoding direction. They occur due to movement mostly and are often seen on the sort of image shown in question 9 in which the popliteal artery appears both above and below its true position. (This is not shown on the image in this exam.)

Case 10

Transcranial ultrasound. Neonatal brain.

1. Right Sylvian fissure
2. Right cingulate gyrus
3. Corpus callosum
4. Left lateral ventricle
5. Third ventricle

The fontanelles are used as acoustic windows to image the neonatal brain. This image is obtained with the probe at the anterior fontanelle and angled a little posteriorly.

Case 11

MRI pelvis (male). T2W sagittal section.

1. Rectum
2. Rectovesical pouch
3. Corpus cavernosum
4. Testis
5. L5/S1 intervertebral disc

Case 12

MRI shoulder. T1W coronal section.

1. Suprascapular vein
2. Acromion
3. Deltoid muscle
4. Long head of biceps tendon
5. Supraspinatus muscle

Case 13

Ultrasound neck. Transverse section.

1. Left lobe of thyroid gland
2. Thyroid isthmus
3. Trachea
4. Left common carotid artery
5. Left sternocleidomastoid muscle

Case 14

MRI base of skull. T2W axial section.

1. Medulla
2. Right vertebral artery
3. Left cerebellar hemisphere
4. Cisterna magna
5. Right mandibular ramus

Case 15

CT thorax with intravenous contrast. Sagittal reconstruction.

1. Pulmonary trunk
2. Right ventricle
3. Left main bronchus
4. Interventricular septum
5. T12

Case 16

MRI abdomen. T2W axial section.

1. Right lobe of liver
2. Transverse colon
3. Inferior vena cava
4. Left renal vein
5. Left crus of the diaphragm

Case 17

CT abdomen with intravenous contrast. Parasagittal reconstruction.

1. S1 vertebral body
2. Duodenum
3. Lower lobe of right lung
4. Inferior vena cava
5. Right renal artery

'D' is clearly the IVC as this passes into the liver superiorly so we are looking at a slightly off-centre parasagittal image just right of the midline. 'E', therefore, is a vessel passing behind the IVC on the right so is the right renal artery.

Case 18

Cardiac MR. T2W long-axis.

1. Left ventricle
2. Left atrium
3. Left pulmonary vein
4. Aortic arch
5. Left hemidiaphragm

Cardiac MRI images are obtained at oblique angles, in line with the axes of the heart and tend to be named as 'long axis' and 'short axis' to match the terminology used in echocardiography.

Case 19

CT thorax with intravenous contrast. Axial section.

1. Superior vena cava
2. Azygos vein
3. Trapezius muscle
4. Right internal mammary vein
5. Left internal mammary artery

As the azygos arch courses anteriorly, oblique sections may make it appear as an ovoid structure on some images. This risks being mistaken for a lymph node by some but the radiologist knows better!

Case 20

MR thigh. STIR axial image.

1. Tibia
2. Fibula
3. Popliteus muscle
4. Popliteal vein
5. Tibialis anterior muscle

Note that the popliteal artery is circular and has a thicker wall than the popliteal vein.

AP	Anteroposterior
CSF	Cerebrospinal fluid
CT	Computed tomography
CTPA	Computed tomographic pulmonary angiogram
DP	Dorsopalmar/dorsoplantar
ERCP	Endoscopic retrograde cholangiopancreatography
IVC	Inferior vena cava
MRI	Magnetic resonance imaging
MRCP	Magnetic resonance cholangiopancreatography
MIP	Maximum intensity projection
PA	Posteroanterior
STIR	Short T1 inversion recovery